The World's Challenge

Marion Guillou • Gérard Matheron

The World's Challenge

Feeding 9 Billion People

Translated by Sarah Hunt, Teri Jones-Villeneuve, Inge Laino and Christine Young

Springer

Marion Guillou
Agreemium, Paris
France

Gérard Matheron
CIRAD
84750 Saint-Martin-de-Castillon
France

Éditions Quæ, R10, 78026 Versailles cedex, France

ISBN 978-94-017-8568-6 ISBN 978-94-017-8569-3 (eBook)
DOI 10.1007/978-94-017-8569-3
Springer Heidelberg New York Dordrecht London

Library of Congress Control Number: 2014931847

Printed on acid-free paper

Springer is part of Springer Science+Business Media (www.springer.com)

Introduction

How many civilisations have collapsed due to a lack of food security[1], or, more precisely, to agricultural systems that fail to satisfy the needs of populations in a context of over-exploitation of natural resources or irreversible environmental damage [1, 2]? The Sumerians, one of the earliest civilisations to practice agriculture some 4,000 years before Christ, used irrigation to develop a productive agricultural system and ensure a surplus of food, thereby giving birth to the world's first cities. Based on a sophisticated social organisation, the irrigation system of Sumerian city-states had one flaw that is believed to have led to the disappearance of this "hydraulic civilisation": irrigation water from dams built along the Euphrates was distributed by a system of gravity canals, allowing only a portion of the water to be used for irrigation while the remainder evaporated or seeped into the ground. Percolation caused water tables to rise and, when it eventually reached surface levels, to evaporate, leading to an increase in soil salinity, compromising soil fertility and resulting in a fatal decline in agricultural yields.

Likewise, the Maya civilisation of the lowlands of Central America is believed to have died out over the course of approximately two centuries, between 750 and 950 AD[2], due to overexploitation of the poor lateritic soils of the southern Yucatan, deforestation and erosion, which severely compromised the soil fertility and led to inadequate agricultural output to meet the nutritional needs of society. An ecological disaster triggered by irreversible environmental damage due to deforestation may also be at the root of the fall of the civilisation of Easter Island. Fierce competition between the tribes that inhabited this far-off Pacific island is believed to be linked to the proliferation of increasingly large statues, the transport of which from stone quarries to the coasts where they were erected required ever-greater quantities of wood. This systematic deforestation could have led to a severe shortage of wood for digging canoes and for birds—another source of nutrition—to build nests.

[1] Numbers in brackets refer to bibliographic references presented by Chapter at the end of the book.

[2] The so-called "Classical" period of the Maya civilisation began around 250 AD and ended approximately 900 AD.

Many scholars, inspired by the ideas of Thomas Robert Malthus or the conclusions of the Meadows Report, commissioned by the Club of Rome in 1973, plead for limiting growth, and sometimes even for a voluntary slow-down. In the same year, the film *Soylent Green* was released. Based on a science fiction novel published in 1966, it was set in an apocalyptic, over-polluted world where both flora and fauna have vanished. Arable land becomes a rare commodity and there is such a shortage of resources to produce already scant "natural" food that man has to resort to synthetic food, including "soylent green", produced by an eponymous multinational. This food product is derived from human cadavers, within the framework of a policy of organised euthanasia of the elderly. In contrast, history is not lacking in examples of civilisations that managed resources successfully over long periods of time, reconciling collective and individual aspirations. The diversity of methods used is worthy of note. Some civilisations, such as the Roman and British Empires, developed or obtained necessary food resources by direct geographical expansion or colonisation. Others achieved food security through sound resource management. Icelandic civilisations, which initially borrowed livestock farming systems from Norway and the British Isles, are a case in point. Aware of the irreversible risks of erosion provoked by over-grazing in high-altitudes, these civilisations took the wise step, in the fourteenth century, of adapting the size of sheep flocks according to the availability of pasture.

This example of coordinated management of a resource, in this case pasture, through organised institutions, is one of three solutions [3]—along with the nationalisation of resources and, on the other extreme, privatisation—to what experts describe as "the tragedy of the commons". This refers to the inevitable rivalry that springs from sharing a common resource, the conflict between private and communal interests, and ultimately the overexploitation of the resource. At the dawn of the twenty-first century, it is precisely this issue, on a global scale, that constitutes the demographic challenge of our planet. How are we to feed the estimated 9 billion inhabitants of Earth by 2050 while preserving ecosystems? What steps must be taken to avoid the overexploitation of resources in the short term? And how are we to implement one or several collective solutions among those planned or deemed feasible that are conducive to the sustainable management of Earth's common resources?

This book explores the status of food security in the world today, but more particularly what is in store for tomorrow. The examples cited above illustrate the need for long-term solutions. Food security is a complex and demanding issue. Complex because it involves more than just producing enough food to feed the world. The 1998 Nobel prize winner in Economic Sciences Amartya Sen wrote as early as 1981: "Starvation is characteristic of some people not having enough to eat. It is not the characteristic of there being not enough to eat." Beyond crises, wars and economic and social inequality—all of which take their toll on food availability, be it permanently or temporarily—available food must be healthy and conducive to a balanced diet. It is therefore not enough to speak solely of "sufficient available supply" or "accessible supply", but of balanced nutritional habits that involve the entire food supply chain, from production to distribution, taking consumers and their environments into account as well. The issue is also demanding because it requires that food be produced without compromising the needs of future generations. In order for this to happen, an assessment must be made of fundamental agricultural practices and food systems, and those that make the most sense in terms of energy, respect for the environment and social justice promoted. Numerous examples have shown that the current food systems of rich countries provide a model that is neither sustainable nor reproducible on a wide scale.

"The economic model of the West cannot be applied to a population of 1.45 billion Chinese people (in 2030)" wrote Lester Brown [4]. Going forward, we must change the way we produce and consume food around the world, and limit waste. The joint ethics committee of INRA (French National Institute for Agricultural Research) and CIRAD (French Agricultural Research Centre for International Development) summed up the stakes of research efforts thus: "It is at once a quantitative and qualitative challenge: quantitative in that global food production must reflect geographic realities, and qualitative in that what is being produced must be nourishing, healthy and affordable to the greatest possible number of people. We must find new ways not only of producing food, but of consuming food, and thereby build a society of moderation where waste is kept in check at every stage of the food supply chain" [5]. To be sure, this is easier said than done. We would do well to admit from the start that satisfying the nutritional needs of a global population of over 9 billion people by 2050 in a sustainable fashion is indeed possible, but only under certain conditions which are outlined in this book. We draw on the lessons learned from past success stories and failures to build a promising future. Among the first and foremost of these conditions are science and innovation. Much more than a simple belief, this is a deep-seated conviction that we as heads of INRA and CIRAD share.

Malthusianism has been the subject of great controversy. Nevertheless, as Daniel Cohen has pointed out, it has withstood the test of time, at least up until the industrial revolution [1]. According to this law, "every time a society begins to prosper, an inevitable mechanism is set in motion to annul the benefits". Economic growth leads to demographic growth as life expectancy increases and mortality rates decline. As populations grow, revenue per capita decreases and it is not long before the population is faced with an inevitable paucity of cultivatable land. "Since there are too many people, some must die, either of starvation or disease" [1]. Up until very recently, that is, the beginning of the nineteenth century, history seemed to confirm this "sinister science": average life expectancy in Louis XIV's France or Queen Victoria's England was not much different from that of hunters and gatherers or the first farming cultures [6]. And this despite the fact that agricultural systems have changed drastically over the past few decades. Whether it be the green revolution in Asia, agricultural progress in Europe or American expansion, technical advances and dedicated socio-economic organisations multiplied agricultural productivity by an average factor of two at the global level between 1960 and 2000. This was accompanied by a significant increase in the productivity gap between different types of agricultural systems [7].

The drawbacks of this unprecedented revolution are well known: environmental, social and economic repercussions, as illustrated by environmental damage and the 2008 food riots. Collective progress depends on a society that is open not only to technological but social innovation, taking all of these issues into account. Swift, consistent action must be taken at all levels, and time is of the essence in predicting and offsetting trends that are already underway. Emmanuel Le Roy Ladurie has noted that the challenge of food supply, food shortages and crises have often set the scene for revolution. Beyond food, it is the very future of human societies that is at stake.

References

1. Cohen D (2009) La prospérité du vice, une introduction (inquiète) à l'économie. Albin Michel, Paris
2. Diamond J (2006) Effondrement. Gallimard, Paris
3. Harding G (1968) The tragedy of the commons, Science
4. Brown L (2007) Le Plan B: pour un pacte écologique mondial. Calmann-Lévy, Paris
5. Inra & Cirad (2009) Avis du comité consultatif commun d'éthique de l'Inra et du Cirad de novembre 2009
6. Clark G (2007) Farewell to Alm. Cité par D. Cohen (2009) La prospérité du vice, une introduction (inquiète) à l'économie. Albin Michel, Paris
7. Mazoyer M, Roudart L (2002) Histoire des agricultures du monde. Seuil, collection Points histoire, Paris

Acknowledgements

The undertaking of such an ambitious project as writing this book would never have been possible without the valuable input and sound advice of Hervé Guyomard, the insight of Etienne Hainzelin, the support of Sylvie Colleu, the objectivity of Patrick Caron, the experience of Pierre Stengel and the ecologist's point of view provided by Laurent Lapchin.

We would also like to thank Perrine Burnod, Benoît Daviron, Paul Colonna, Driss Ezzine de Blas, Jean-Christophe Glaszmann, Christian Huyghe, Barbara Redlingshöfer, Anaïs Tibi and Valérie Toureau, whose invaluable contributions greatly enriched our approach.

Our wholehearted thanks go out to each and every one of them.

Contents

Chapter 1
Setting the Stage

If the course of history has been marked by famine and food shortages, it was thought that the twentieth century provided the technical, social and economic solutions to make such scourges a thing of the past. The proportion of the global population suffering from undernourishment decreased slowly but steadily and international organisations set ambitious targets for world leaders in terms of tackling poverty and world hunger [1]. Nevertheless, some claim that this relative food security will be threatened in coming years, both in the short and long term [2, 3].

Since Neolithic times, man has depended on agricultural activity and fishing to provide essential foodstuffs and the necessary raw materials to meet other needs. The frenzied race of colonial powers towards islands rich in sugar and spices and lands producing cotton and rubber is a case in point. The twentieth century made great strides in terms of the mechanical and chemical industrialisation of agriculture, followed by the development of the service industry, leading the world to believe that agriculture could be relegated to a secondary activity, one unlikely to generate significant wealth or drive economic growth. Simultaneously, in wealthy nations, urbanisation and new means of production further weakened the connection between food and agriculture, undermining agriculture's former quasi-sacred role as a food-providing process and reducing it to a purely economic activity. The Washington Consensus of the early 1980s served to definitively sever whatever remained of that link. It is now widely recognised that hunger is not solely an agricultural issue but has as much to do with access to food, that is, the solvency of an ever-more urban consumer population. Indeed, the issue of hunger is inextricably linked to development issues and reducing world poverty [4].

The World Bank's 2007 development report [5] draws lessons from the mistakes made based on the false notion that agriculture can be reduced to a secondary activity. It sends out a loud and clear message about the key role agriculture plays as "an essential tool for achieving the Millennium Development Goals, which consist of halving by 2015 the proportion of the global population suffering from extreme poverty and chronic hunger". At the same time, the United Kingdom, although traditionally firm in its conviction that international exchange ensures world food security, launched the Global Food and Farming Futures project [6], a large-scale

M. Guillou, G. Matheron, *The World's Challenge*,
DOI 10.1007/978-94-017-8569-3_1, © Éditions Quæ, 2014

forecast of the global agriculture and food system. The 2008 food riots were a rally cry for world leaders to step up and ensure the basic human right of food security[1].

Science Enters the Equation

In 2004, scientists were called upon for the first time to take stock of agriculture-related achievements and outlooks, and provide an overview of the situation based on the participative approach used in climate change. They were entrusted with identifying how agricultural know-how, sciences and technology can contribute to development and sustainability while boosting agricultural yields. Some four hundred experts were consulted. They produced an in-depth and unprecedented analytical report providing not only technological data but also clearly revealing the need to rethink public policies on the issue, organisational capacities and the role institutions can play [7]. National scientific institutions got involved as well: in the UK, the Royal Society took up the issue and provided an analysis of the situation in 2009 before calling for widespread action [8].

In 2006, INRA and CIRAD joined forces to launch a foresight study on global food and agricultural systems and related issues for 2050 called Agrimonde [9]. The stakes are clear yet ambitious: ensure food security for a global population of some 9 billion people by 2050 while preserving the planet. The goal in launching the study was to make forecasts of possible future scenarios in a highly uncertain economic, social and environmental global context. Although the approach is primarily quantitative and cumulative, grouping countries into broad geographical zones, the study served to spur thinking, open up a forum for discussion with many key players, inform the decision-making process and provide the opportunity to participate in international debates with experts in the field.

A Closer Look at the Issue

The first factor to consider is a growing global population, which doubled between 1961 and 2003, from 3 to 6 billion people. Global food availability per person likewise increased, from slightly less than 2,500 kcal per day in 1961 to slightly more than 3,000 in 2003. Nevertheless, there are still wide disparities throughout the world, which was precisely at the root of the 2008 food riots. Average daily food availability currently stands at approximately 4,000 kcal in industrialised OECD[2] countries, but only 2,500 in many parts of the world such as sub-Saharan Africa and Asia. Moreover, animal products (meat, dairy, eggs, etc.) represent 30 % of calorie intake in Northern countries, but less than 6 % in sub-Saharan Africa. In

[1] The Universal Declaration of Human Rights, 10 December 1948.

[2] Organisation for Economic Co-operation and Development.

order to fully grasp the implications of such disparities, it is worth recalling that the FAO[3] considers satisfactory food availability to be 3,000 kcal per day per person. Further on, we will explore the difference between food availability and actual consumption.

From a quantitative point of view therefore, the planet is currently able, on average, to nourish the global population with an overall food availability of 3,000 kcal per day per person. The merit of the Agrimonde study lies in the fact that it goes beyond average figures, stretching the horizon to 2050 and factoring in the diverse situations of different populations in different regions of the world, and uses this timeframe in order to provide research-based solutions and redress imbalances. To carry out the study and take a wide range of variables into account including geopolitical, social, cultural, sanitary, economic, agronomic, ecologic and technological factors, Agrimonde uses an original quantitative tool: Agribiom. This tool uses a single unit, known as the kilocalorie—sometimes called a food calorie—to measure the whole of agricultural output, exchanges and the use of food biomass in all regions of the world from 1961 to 2003. Data exceeding the 30 million mark were gathered in 51 countries or groups of countries for a period of over 43 years, for the most part from long-term FAO series studies. These were gathered, sorted and used to set up models. Agribiom was used not only to carry out retrospective analyses covering the period from 1961 to 2003, but also to conduct foresight studies.

The advantage of expressing all foodstuffs, whether of plant or animal origin, in terms of calories is that it allows for a simplified comparison of production and consumption systems. It is also one of the distinguishing features of the study. This simplification nevertheless has its drawbacks, notably when it comes to nutrition. The use of the kilocalorie as a unifying factor does not take into account the proportion of macronutrients made up of carbohydrates, proteins and lipids necessary for a balanced diet, and fails to recognise micronutrients such as vitamins and trace elements.

Why Nine Billion?

The foresight study Agrimonde is based on median UN forecasts of population trends in different regions of the world. The global population in 2050 was estimated at 9 billion but forecasts range from 7 to 11 billion. Expected migration trends are relatively moderate, involving around 100 million people, which the UN considers "normal" for cross-border migration over a 50-year period. Within the framework of the Agrimonde study, the underlying principle is to create scenarios that reveal whether or not a global population of 9 billion by 2050 will be able to depend on an adequate food supply. The two scenarios presented reach this goal, but in very different ways.

[3] United Nations Food and Agriculture Organization.

Food for all: Two Different Scenarios

The first scenario, Agrimonde GO (for Global Orchestration), is based on established trends. In this context, the 2050 global population will be fed thanks to economic growth and innovation. Set goals include increased employment, material comfort, a healthy population and scientific and technological progress. This scenario projects historical trends into the future within the context of a liberalised world. Food consumption and production both increase, notably through greater yields, comparable to those recorded since the postwar years. Food availability increases in all regions of the world, reaching an average 3,590 kcal per day per person. Under-nourishment is considerably curtailed. Nevertheless, wide disparities between geographic regions persist, e.g. 3,000 kcal in sub-Saharan Africa compared with 4,000 kcal in developed countries. What is more, this scenario is destined to result in environmental degradation.

Insert 1.1. Produced Calories or Consumed Calories: What is the Difference [10, 11]?

There is a big difference between calories produced in the field and calories actually consumed by people. A bit of clarification is in order to provide valid estimates and accurate comparisons.

The quantity of calories produced by farmers is what is referred to as food availability. In order to simplify matters, food availability is expressed in calories produced per person per day. At the end of the 1990s, average global food availability was estimated at 4,600 kcal per person per day. At national or multinational level, food availability is calculated based on domestic production plus imports, minus exports, and plus stock levels (positive if stock increases, negative if stock decreases). Losses incurred during harvest, estimated at 600 kcal per person per day, are then subtracted from this figure to determine apparent food availability, which stands at 4,000 kcal per person per day. These 4,000 kcal do not, however, find their way onto people's plates! Some 1,700 of them are used to feed animals, which in turn yield 500 kcal in animal-based foodstuffs—and some further 800 kcal are lost or wasted during processing, from distribution to final consumption, in or outside the home. In the end, only 2,000 kcal per person per day are actually ingested by humans. This is what is referred to as food consumption.

Between the initial 4,600 kcal and the 2,000 ultimately consumed, the journey from farm to fork is riddled with obstacles: 57 % of produced calories do not make it to the intended end use! This figure gives pause for thought in a world where close to 1 billion people are undernourished, and whose population will continue to grow, especially in developing countries. Finding ways to spur progress is a key challenge of research.

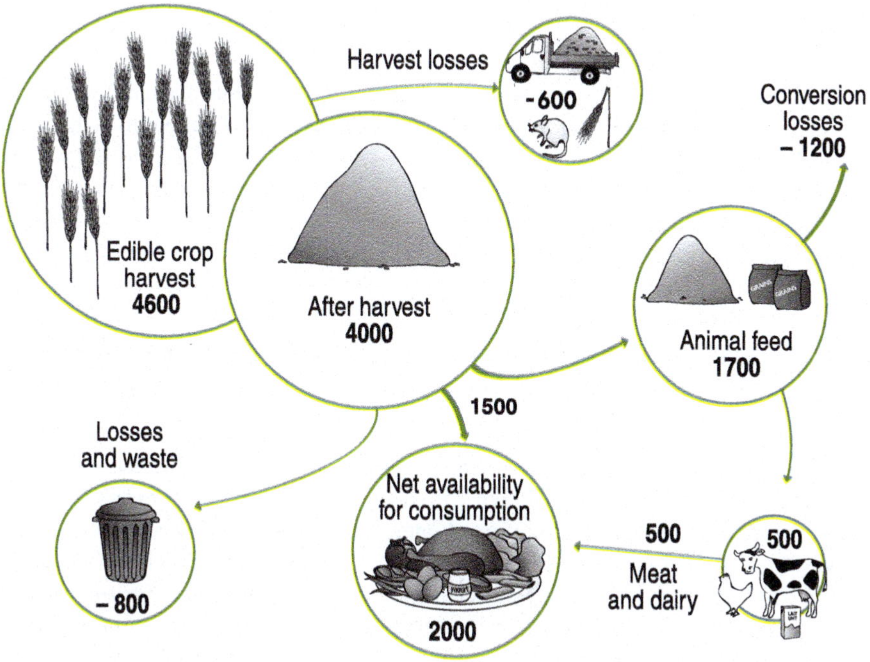

Fig. 1.1 From farm to fork, the journey of a kilocalorie (in kilocalories per person per day). (Source: Smil V. 2000. [10, 11], reproduced with the permission of F. Bourin Publishers.)

The second scenario, Agrimonde 1, is pro-active, breaking away from current trends. In this scenario, the world can be fed thanks to an agricultural system and food industry which are sustainable on three fronts—economic, societal and environmental—and which are based on production technologies that preserve ecosystems. This scenario allows for economic and social development through agriculture in countries that do not have the means to implement production systems requiring significant investments in terms of equipment, water, fertilisers, pesticides, etc. When it comes to consumption, this scenario tends to reduce inequalities related to nutrition and health by curbing both under-nourishment and overnourishment. Agrimonde 1 is a normative scenario which sets the goal of average food availability at 3,000 kcal per day per person in all regions of the world, all the while taking into account differences in the nature of animal-based calories consumed (pork, poultry, beef and fish). This even distribution of food availability throughout the world is in sharp contrast to trends observed over the past decades. It is also in contrast with established eating habits, notably in western nations: the average 3,000 kcal break down into 2,500 kcal derived from plant products and 500 from animal products (meat, eggs, milk and dairy products, etc.), i.e. 20 % of calorie intake instead of the current 30 % typical of western nations today (Fig. 1.1).

Agrimonde: The Results

The first and foremost lesson learned from the two different scenarios is that it is indeed quantitatively feasible to feed the entire planet. Compared with 2003, global agriculture yields increase by 88 % with Agrimonde GO and 28 % with Agrimonde 1. In both cases, this is less than double the current production. The following pages take a look at how each scenario reaches its goals, their distinguishing characteristics, and respective advantages and disadvantages.

Agrimonde GO: Feeding the World Through Economic Growth in a Context Where the Environment is not the Top Priority

In the Agrimonde GO scenario, economic growth is robust in all regions of the world, notably in Asia, sub-Saharan Africa and former Soviet countries thanks to two factors: liberal trade and reinforced international economic cooperation. Heavy investments in research, innovation, education, health and infrastructure, in particular in developing nations, and significant technical advances, are the norm. Poverty is substantially reduced. Agricultural production rises significantly in all geographic zones, thanks primarily to increased yields, but also to increased cultivated land (+330 million hectares between 2003 and 2050) and grazed pastures (+260 million hectares during the same period). Deforestation rates are nevertheless much lower than those recorded from 1960 to 2000. Agricultural production systems are increasingly homogeneous and in line with a benchmark industrial model which relies heavily on motorisation and mechanisation, mineral fertilisers and pesticides. The distinguishing features and specific resources of areas are not factored in, and diversity is compromised, particularly in terms of cultivated plant species. Irrigation is key to boosting yields.

Higher agricultural output ensures that sharply increased food requirements are met, in a context of evolving demographics and overall improved quality of life. The number of undernourished is divided by 2.5 within 50 years. All over the world, the portion of animal-based calories increases. Global economic growth goes hand in hand with growing energy needs, which are met thanks to technical advances and investments. Approximately 10 % of global electricity is generated from renewable energies or biomass. The latter is also used to produce agro-fuel on a wide scale, creating a significant need for exploitable land. Land used for non-food related purposes increases six-fold over 50 years.

This scenario is hardly conducive to solving global environmental issues; on the contrary, climate change and the over-exploitation of marine resources (and renewable resources in general) are exacerbated and solutions to offset them are not provided. Environmental concerns take a back seat while priority is given to satisfying immediate dietary and energy requirements. Solutions to environmental problems are implemented on a purely case by case basis. Despite the availability of financial resources and competent global institutions, no real effort is made to cut greenhouse gas (GHG) emissions or preserve biodiversity.

Agrimonde 1: Feeding the World in a Context of Sustainable Development

Agrimonde 1 provides a completely different scenario. Under the joint effects of accelerated climate change and multiple food and energy crises that mark the beginning of the period in question (2010–2020), the global response consists of adopting sustainable development methods. Global economic growth is driven by developing nations, who rely primarily on the prosperity and expansion of rural areas. Development is part of a cumulative, virtuous circle. Thanks to international aid and national financial transactions, initial investments are secured, in particular in terms of storage infrastructure, transport, preliminary industrial processing, education, health, etc. This leads to increased household incomes in rural areas which in turn opens doors to new investments. Increased household incomes in rural areas of developing countries also leads to improved diets in terms of quantity and diversity.

More precisely, food availability in 2050 stands at 3,000 kacl per day per person in all regions of the world, 2,500 of which are of plant origin. This average overall figure belies discrepancies between regional trends from now until 2050. In sub-Saharan Africa, food availability increases considerably thanks to agricultural development. In contrast, it decreases by nearly 25 % in developed OECD countries under a two-fold influence: on the one hand, reduced losses and waste during distribution and final consumption and, on the other hand, more efficient food policies. Diets are more varied thanks to several factors including culture and food policies.

Insert 1.2. What Role Will Animal Products Play in Tomorrow's World?

The Agrimonde foresight study sheds light on the complexity of issues related to animal products and livestock farming. In the Agrimonde 1 scenario, the global population can be fed by establishing food availability at 3,000 kcal across the board. Some 500 of those are of animal origin. However, this is quite a big assumption, since history has shown that as incomes increase and populations become more urbanised, the consumption of animal products likewise increases until it reaches a plateau. Agrimonde 1 strives therefore to limit the consumption of animal products in developed countries and reverse current trends that are taking root in developing countries as well as in the most well-off households of poor nations.

What is the point of focusing on a small portion of animal produce in developed and developing countries? The answer lies in what is known as the "plant biomass to animal biomass conversion rate". The Agrimonde foresight study is based on the estimate that between two and five plant-based calories are needed, depending on geographical region, to produce one animal-based calorie. The consumption of animal-based calories therefore seems less energy efficient. Reducing the share of animal-based calories in diets would allow lands currently used for animal husbandry to be given over to plant production. However, care must be taken not to oversimplify matters. Animal husbandry brings a plethora of goods and services to mankind: animal traction, transport, manure to fertilise land, wool, leather, landscape maintenance, the preservation of biodiversity through the maintenance and management of grasslands, etc. Not to mention that ruminants, not men, feed on grass and scrub, and that 1 billion people in the world depend on animal husbandry to make a living. This activity is therefore an essential component of a sustainable global agricultural system.

Growth in agricultural production, on which the global food supply depends, is significantly lower in Agrimonde 1 than in Agrimonde GO. Overall output increases, but at very different rates depending on the region: slight increases in OECD countries, Asia, North Africa and the Middle East and significant increases in Latin America and sub-Saharan Africa. Cultivated land increases considerably—approximately 580 million hectares over 50 years, i.e. a two-fold increase from Agrimonde GO—to the almost exclusive detriment of grazing. Forest cover remains practically unchanged. Increased yields, less pronounced on average than in Agrimonde GO, depend on multiple factors, constituting a major difference between the two scenarios. Innovation abounds, springing from local know-how and ecosystem services, all the while giving preference to technologies that favour environmentally friendly agricultural practices. Greenhouse gas emissions from agricultural activity are reduced, soil and water resources better protected, and biodiversity preserved and enhanced. Nevertheless, local agricultural production remains insufficient to meet domestic demand in North Africa, the Middle East, sub-Saharan Africa and Asia. The shortfall is compensated and food security ensured in each zone by imports from the other three zones, within the framework of secured and regulated international exchange under the aegis of the UN. International regulations allow for significant departures from established liberal exchange policies to take environmental risk into account and protect local agricultures, which, in the developmental stages, are less productive.

Lessons Learned from the Agrimonde Study: Change is in Order

The point of providing two scenarios as different as Agrimonde GO and Agrimonde 1 is to highlight the key issues at hand and reveal what changes are called for. The first driver of change consists in reducing excessive food consumption in rich countries and well-off households. A similar trend has been observed in industrialised countries where change has been well underway for decades, in emerging economies where change is more recent but occurs much faster, and among the upper classes and urban populations of poor nations: as the need for calories decreases, consumption increases. This imbalance results in a sharp increase in obesity or overweight, but also in qualitative food imbalances, a major public health challenge. Chapter 2 takes a closer look at this issue.

In wealthy nations, approximately one third of food is lost or wasted in industrial processing, distribution and final consumption. All players, both public and private, must make reducing losses and waste a priority. All the more so because 1 kg of wasted food represents a loss in terms of all the components that went into producing that food (land, water, energy, fertiliser, pesticides, etc.), not to mention the onus of treating the waste itself. This issue is explored in Chap. 3. Nevertheless, cutting back on excessive food consumption and reducing waste is not enough; food availability must also be boosted. Changes must be implemented early on in the process: cutting post-harvest losses, which are particularly heavy in Southern countries, especially during initial storage and transport (Chap. 4), is a logical next step perfectly in line with reducing losses and waste at the processing, distribution and consumption stages. It is not enough, however, to cut losses in the field and during harvest, especially since goods other than foodstuffs must be produced, such as bioenergies and bioproducts, to compensate for dwindling fossil energy resources (Chap. 5). Moreover, close attention must be paid to what are known as ecosystem services, that is, services provided by nature, including the setting in which history unfolds and societies are organised. A key challenge of the twenty-first century is indeed to find agricultural solutions that respect ecosystems and put them to good use by implementing integrated, energy-smart management strategies that limit external expenditure on farm inputs and take local situations and societal priorities into account (Chap. 6).

The question of boosting agricultural production remains. How can this be achieved? Is it possible to dedicate more land to farming? Yes, but this driver of change cannot take top priority, even if uncultivated arable land is indeed available. This land cannot be used for production on any significant or sustainable scale as it constitutes carbon sinks and provides havens for biodiversity. Chapter 7 addresses the issue of competing land uses. As in the past, boosting agricultural yield will be key, while respecting the environment and natural resources. The ultimate goal: an agricultural system that is at once productive and ecological. How can this be achieved? Chapter 8 provides some answers.

The drivers of the changes mentioned above all have their merit. Nevertheless, they alone will not suffice to solve the problems of undernourishment and access to a quantitatively and qualitatively adequate diet. These issues are primarily related to poverty, and therefore to economic development (Chap. 9). Food autonomy is but a pipe dream in some regions of the world. Even if food-producing agriculture were to develop rapidly in developing countries, it would still not be enough to meet the needs of local populations. Regional and international trade in agricultural and food products will be even more necessary than it is today. In light of this, countries must come together to protect and promote trade, ensure food security for their populations, especially in times of crisis, and stabilise excessively volatile market prices, which in turn impact food prices. More than ever, it is believed that this will have to be based on concerted governance on a global scale. This will be explored in further detail in Chap. 10.

The international scientific community has sounded the alarm: the agricultural system that will feed the world in 2050 will not be the agricultural system in use today. Innovation on a broad scale is in order, but all the evidence points to the fact that it is absolutely necessary to invest in agriculture, the food industry and food supply, in rich and poor countries alike. A child born today will live on a planet shaped by the choices made today. Now is the time to step up and face the issues, and gear actions toward ensuring a viable future for the planet's population. Earth can feed 9 billion people in 2050, provided the importance of agriculture is rediscovered.

References

1. Objectifs du Millénaire pour le développement, rapport 2010, Nations unies, New York, juin 2010
2. Von Braun J (2007) The world food situation: new driving forces and required actions. IFPRI, Washington D.C.
3. Brown L (2007) Le Plan B: pour un pacte écologique mondial. Calmann-Lévy, Paris
4. Sen A (1982) Poverty and famines: an essay on entitlements and deprivation. Clarendon Press, Oxford
5. Banque mondiale (2007) Rapport 2008 sur le développement dans le monde: l'agriculture au service du développement. Banque mondiale, Washington D.C. (Published on 19/10/2007)
6. UK Government, Department for Business, Innovation and Skills (2011) Global food and farming futures. http://www.bis.gov.uk/foresight/our-work/projects/current-projects/global-food-and-farming-futures. Accessed Sept 2011
7. IAASTD (International Assessment of Agricultural Knowledge, Science and Technology for Development) (2005) Agriculture at a crossroads, rapport global, 2009 (a sommary exists in several languages for decision-makers)
8. The Royal Society (2009) Reaping the benefits. Royal Society, Londres
9. Paillard S, Treyer S, Dorin B (coordinateurs) (2010) Agrimonde: scénarios et défis pour nourrir le monde en 2050. Éditions QUAE, Paris
10. Lundqvist J, de Fraiture C, Molden D (2008) Saving water: from field to fork. SIWI Policy Brief. From [11]
11. D'après Smil V (2000) Feeding the world. The MIT Press

Chapter 2
Eat Well, Eat Better

For millions of years, man's primary preoccupation was having food to eat. Today, eating is an act that goes far beyond fulfilling biological needs, touching upon the social, hedonistic, cultural and symbolic aspects of life. Malnutrition includes illnesses related to overeating, undereating, or dietary imbalances in terms of food quality. These imbalances, most often linked to lifestyle, are on the rise throughout the world. Eating habits depend on much more than just individual preferences or consumer purchasing power, as was pointed out in the collective scientific expert report on eating habits coordinated by INRA in 2010. They are a determining factor not only for personal health but also for the diversity of the world's food systems.

Throughout history, human societies have experienced changes in lifestyle and quality of life as well as in the availability and nature of food. When the changes involved are particularly profound, these are known as "food transitions". An example would be the changeover in Neolithic times from hunting and gathering to farming and animal husbandry, or the passage from times of famine to times of relative prosperity such as developed countries are experiencing today. The food transition that developed countries have been experiencing for the past 100 years is characterised by a growing imbalance between the abundance of food products, their relatively easy access, and an increasingly sedentary lifestyle leading to a decreased need for calories. This imbalance in turn triggers a slew of health problems, from obesity to chronic diet-related illness such as diabetes, cardiovascular disease or articular disease. These food transitions also go hand in hand with epidemiological changes.

Today, some population segments in many developing countries are experiencing similar food transitions, combining excessive calorie intake and an increasingly sedentary lifestyle. What is more, this transition is taking place in relatively short time spans. Even if the issue of hunger stubbornly persists, especially in developing countries, increasingly urbanised populations in emerging economies, and even in poor countries, are no stranger to illnesses linked to overeating. In fact, analyses on a country-by-country basis are of little interest: just as there are people who go hungry in rich nations, there is a growing number of people who suffer from dietary imbalances in all countries around the world. The key public health challenge of the future will no longer be infectious diseases, but so-called degenerative diseases. To redress these imbalances, which depend on multiple and complex factors,

M. Guillou, G. Matheron, *The World's Challenge*,
DOI 10.1007/978-94-017-8569-3_2, © Éditions Quæ, 2014

well-thought-out action must be taken at all stages of the supply chain (agricultural products, industrial processing, distribution, catering services, etc.), as well as at demand-side (final consumers, consumer preferences and habits, constraints, ability to make choices, etc.). Regulatory strategies must also be taken into account (industry regulations, informing consumers and raising awareness, tax programmes or subsidies, etc.). This chapter presents a series of actions pertaining to different stages of the food system that come into play once the produce leaves the farm. Particular attention will be paid to consumers, distributors, players in the food industry, and the authorities entrusted with public health issues. If sustainable solutions are to be found, the perspective taken must be a global one, acknowledging the role lifestyle plays. This is the new food transition nutritionists are calling for.

Changes at the Root of Nutritional Imbalances

Food transitions that have long been underway in wealthy nations, and more recently but also more rapidly in emerging economies and well-to-do households of poor nations, are most often characterised by a rise in calorie consumption despite a decline in calorie requirements. This imbalance results in a surge of health issues such as overweight and obesity, and qualitative nutritional imbalances.

Changing Food Consumption Habits in the Developed World: France, a Case in Point

Contents of the Consumer Shopping Basket have Changed

In terms of food availability, the twentieth century is marked by profound changes. Energy rations reached a plateau and then gave way to a phase of qualitative changes and diversification of diets [1]. Figure 2.1 illustrates this trend in France. Graph A shows the evolution of total available calories per person per day between the end of the eighteenth century and 1960, as well as changes in the principal food sources of these calories (plant *vs* animal origin). Graph B provides a breakdown of total calorie intake by macronutrients (carbohydrates, lipids and proteins) between 1780 and 2000.

Trends in apparent food availability are related to documented trends in food consumption. The share of cereals (bread, pasta) in the food ration has been seen to decrease, followed by a drop in starches (potatoes), while simultaneously more meat products, fruits and vegetables, fatty foods and sugar are consumed. This phase continues up until the 1980s before levelling off, with a slight drop in calorie intake. Such a trend is not unique to France but has been observed in all countries of the Mediterranean basin, where a significant rise in available calories and animal-based lipids, and a drop in carbohydrates, have been recorded. The same phenomena are currently being observed in Northern and Eastern Europe [2]. When

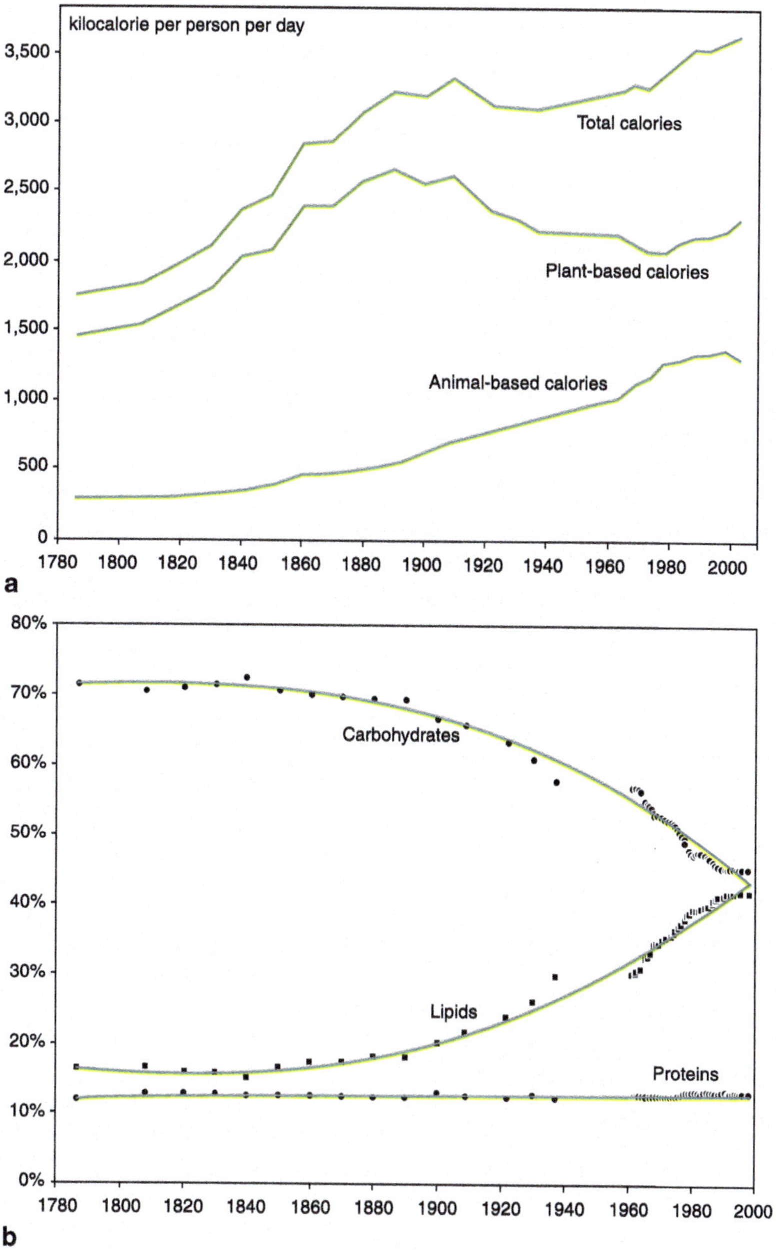

Fig. 2.1 Level of apparent food availability in France. (Source: P. Combris, based on J. C. Toutain, FAO Stat, reproduced with the permission of F. Bourin Publishers)

it comes to meat products, the trend is less linear. It can nevertheless be described as an overall drop in consumption from the 1980s onward, a shift from butcher meats to poultry, steady pork consumption, and an increase in fish intake. Some foods formerly in low demand have witnessed an exponential rise in consumption: dairy products (cheese and yogurt), fruit and vegetable juices, and frozen or processed foods [3].

Other products are also characterised by less linear change: fresh fruits and vegetables remain stable in terms of volume [4] while processed fruits and vegetables are clearly on the rise [5]. One of the major trends in the past decades has indeed been a shift in consumption toward ever-more processed food [6]. Among industrial food products, those in highest demand are those that are the most accessible from an economic point of view, thanks to economies of scale, with large-volume production and the development of wide-scale distribution [7]. The relative stability of the share of industrial food products[1] in household food budgets (80 % in 1960 and 83 % in 2006) goes alongside a rise in the volume of food consumed. This overall trend holds true for the most heavily processed food products [5] which require minimum preparation time. The boom in fresh dairy products (25-fold increase in volume consumed over 40 years) is the most telling example of the shift from basic products to more heavily processed products. Canned goods and ready meals also increased in popularity, on the supply- as well as demand-side: annual household spending went from 639 € in 1979 to 872 € in 2000 [8]. All of these trends taken together lead to an increase in the "calorie density" of food purchased and consumed.

Large Supermarket Chains Now Play a Key Role in Household Food Supply

Although homegrown consumption is still widespread in France (25 % of households have a vegetable garden and 15 % an orchard), this is nevertheless limited to certain products (fruits and vegetables, rabbits and poultry, etc.). Today, household food supply still depends on where food is obtained, which in turn depends on consumer profile (income, age, etc.), place of residence, proximity to a supermarket or hypermarket, etc. This notwithstanding, the size and predominance of markets, small retail outlets and large supermarket chains have been changing considerably over the past 40 years. Hypermarkets, which have long held around 5 % of market share, accounted for a third of the share at the close of the twentieth century. Since then, that figure has remained stable or dipped slightly. Supermarkets have witnessed similar margins of growth, today accounting for 70 % of spending on food (compared with less than 5 % in 1970). Simultaneously, the market share of small retail outlets went from 20 % in 1970 to less than 10 % in the early 2000s. Among specialised small food businesses, bakeries have held up best [9–11]. According to surveys on health and nutrition carried out in 2002 and 2008, there seems to be a trend in French households toward diversifying where and how food is obtained. More food is being purchased over the Internet or by telephone, or delivered to the home [12, 13], and direct sales from producer to consumer are on the rise, although still rare [14].

[1] Products from the meat, dairy, beverage and grain industries, as well as "diversified" food industries, as opposed to products directly obtained from agricultural activity (unprocessed fish products, fresh fruit and vegetables, eggs, honey, unprocessed rabbit and game).

Eating In: A Tradition that is Alive and Well

Mealtimes are, without a doubt, an important part of family and community life, revealing the value placed on celebration and pleasure. The fairly widespread notion that this tradition is losing ground in France is difficult to prove. To be sure, profound changes since the 1950s and 1960s have significantly modified traditional mealtime practices in France, due in part to an increasingly urbanised population. Nevertheless, the debate about the breakdown of the traditional French mealtime model, largely under the influence of marketing professionals, continues. French sociologist Nicolas Herpin [15] associates this breakdown with a certain dissolution of meals (simplification of main meals and shift to smaller or side meals), temporal dislocation (imprecise mealtimes), de-synchronising (family members no longer eat together), relocation (multiplication of places where meals are eaten), and a certain de-ritualising (blurring of boundaries between ordinary and festive meals). According to his research, these trends are not equally representative of all strata of French society. More recent studies on inter-generational relations reveal a deregulation of family life, increasingly individualistic lifestyles among family members, and a growing tendency among adolescents to adopt independent lifestyles at a much earlier age [16]. These trends taken together foster greater autonomy when it comes to food and eating habits as well. The relative loss of the importance of family meals combined with the increasing popularity of ready meals, consumed in front of the television or in the living room, sometimes even with guests, bears witness to this.

Nevertheless, the vast majority of the French still have their meals at home. In 1997, four out of five French people had their midday meal during the working week at home, a figure that rises to 90 % for the evening meal [17]. The most recent studies have shown a slight drop in the first statistic while the second remains stable. Today, however, three quarters of the adult population still eat at home at midday, albeit with strong variations depending on social class and place of residence. The budget share of food consumed outside the home among French households stands practically unchanged for the past 50 years, at around 5 %. Of that 5 %, the share spent in restaurants has doubled, from 30 % in 1960 to 60 % in 2005, to the detriment primarily of cafés. The total share spent in public eating establishments has remained stable, at around 23 %.

Growing Similarities Between Food Consumption Models Around the World

Data used to analyse food consumption trends vary from country to country. Studies based on individual habits are the most representative of actual consumer behaviour, but unfortunately, their methodology varies in time and space, making accurate comparisons between countries difficult. In order to evaluate apparent food availability, statistics from the FAO, which include figures for food production and food exports and imports, are most often used. The advantage of this method is that it is homogeneous in time and space, making it possible to compare different countries on a given date and to track trends in each of them. The disadvantage, aside from the varying quality of data collected from different countries, is that the calculation of food quantities actually consumed is far from accurate (see Insert, Chap. 1). Nevertheless, clear trends have been detected.

A rapid food transition has clearly been underway in China for the past 30 years [18]. Great strides have been made in tackling undernutrition and poverty, with annual economic growth rates of 10 %, which have led to greater calorie availability. For the wealthiest Chinese, especially those living in cities, traditional diets characterised by a very small calorie share from lipids, have undergone drastic change. The contribution of calories from meat rose 349 % in the past 40 years, (from 90 to 644 kcal per day per person), that from vegetable-based oils 680 % (from 35 to 273) and that from sugar 305 % (from 18 to 73). In contrast, the contribution of calories from rice only rose 24 %, and that from tubers and dried pulses dropped 31 and 88 % respectively [19].

Today, fats account for more than 30 % of calorie intake among almost half of all Chinese adults [18]. Data have also been collected regarding the calorie requirements of this population, which have dropped as the Chinese practice less physical activity, leisure activities become increasingly passive (e.g. television) and more people have their own cars. Simultaneously, food distribution has also undergone considerable changes, with the proliferation of large supermarket chains and substantial investments. Nevertheless, some distinguishing characteristics remain relatively unchanged, at least thus far: the consumption of non-alcoholic beverages and ready-made meals remains very marginal. As a direct consequence of these shifts, the number of overweight Chinese men and women tripled and doubled, respectively, between 1989 and 2000; between 1985 and 2000, the number of overweight youth rose by a factor of 28. Today, some 290 million Chinese people, i.e. one quarter of the adult population, are overweight, and the number is rising sharply. Over a period of 24 years, the mortality rate due to noncommunicable diseases related to diet has gone up considerably, particularly affecting the underprivileged classes of Chinese society [18].

Trends in food distribution that have been observed in China are also underway in many developing countries, to varying degrees depending on wealth and contact with the outside world. Traditional distribution channels, small and varied, have been replaced with large supermarket chains that are centralised and uniform. The changeover started with staple products (cereals, oil, sugar, etc.) before spreading to fresh produce, albeit with differences from region to region. Large "western-style" supermarket chains attract first and foremost the privileged classes, but quickly draw the urban masses as well, as they are more compatible with urban lifestyles, themselves "westernized": ever-more urbanised societies, bolstered purchasing power, more women working outside of the home, growing number of private vehicles and refrigerators, demand for a wide range of food products, etc.

Practically speaking, this wave of transformations in food distribution swept the world in four phases: first in big cities of the richest nations of Latin America; followed by East and South-East Asia; followed by the poorest Latin American nations and East and Southern Africa, as well as smaller cities of countries from the first phase; and lastly South Asia. The speed of this transformation is impressive. Take Brazil, for instance, where today 70 % of food is distributed in large supermarkets, up from 10 % only 30 years ago. China, Brazil, and many other developing countries have witnessed the same transformation that occurred over a period of 50 years in France and the US in a span of less than 20 years. Incidentally, this rapid development of wide-scale distribution affects the relationships between producers, the food processing industry and consumers. Access to the market becomes more difficult for small producers with limited negotiating clout, or due to stringent specifications in terms of quantity, quality and hygiene. The food purchasing departments of large supermarket chains exert tremendous pressure on producers and industrial players who initiate food processing procedures.

Getting to the Root of Profound Changes

Incomes and Prices that Foster Abundance, Availability and Access

Research carried out in economics and the history of economics has shown that changes in diets are rooted in economic and nutritional factors. Analysed from a broad, overall perspective, that is, based on macronutrients and food groups, these changes are similar from one country to the next, and depend directly on the level of economic development. In industrialised countries, and now in the vast majority of the world, the agricultural revolution, spurred on and followed by the industrial revolution, led to a considerable drop in the cost of food calories. The actual price of different food products changed dramatically, as did diets. The benefits of these changes are well known: biological capacity, aptitude for work, longevity and improved quality of life. The drawbacks, although analysed more recently, merit equal attention: issues related to overweight and obesity, diabetes, cardio-vascular disease, etc.

Research carried out based on FAO surveys allows us to identify and describe nutritional trends that accompany economic development. On the basis of studies carried out at the end of the 1930s in 70 countries, Cépède and Lengellé [20] have shown that people seek to satisfy nutritional requirements by turning first to inexpensive foods such as cereals and tubers, followed by fatty foods, then sugar, and finally meat and dairy products as their quality of life increases. These more expensive food products replace the aforementioned cheaper ones once food needs are met across the board, thus accelerating changes in overall diet.

Périssé, Sizaret and François [21] set out to establish correlations between calorie intake in terms of macronutrients, and per capita income, basing their research on observations made in the early 1960s in 85 countries. They found that as income increases, the share of lipids in diets sharply rises (despite a drop in plant-based lipids); the share of carbohydrates decreases, with a drop in cereals left uncompensated by increased sugar consumption; and protein calories remain stable, as a rise in animal-based proteins compensates exactly for a drop in plant-based proteins. These changes in diet are directly related to the increased consumption of animal products that goes hand in hand with higher incomes.

Eating Habits: The Determining Factors

Sensorial Perceptions, Palatability and Food Intake

When it comes to the consumer, behaviours must be analysed during the food purchasing process and during mealtimes. Although eating is a codified activity governed by social and cultural norms, a certain physiological cycle still comes into play, dividing the process into three phases: a feeling of hunger, the act of eating and a feeling of satiety. Each phase is governed by complex physiological processes, which experts are constantly learning more about.

Experiments have shown that the sensorial properties of food come into play in each of the three phases mentioned above. Although this comes as no surprise, the mechanisms at work are more complex than the simple quantitative satisfaction of a physiological need. Behaviour prior to ingestion is associated with feelings of hunger (physiological need translated into internal signals) and a heightened appetite (desire to eat food that procures pleasure and satisfaction). The initiation of a meal depends, therefore, on factors other than internal hunger signals. Pre-ingestive behaviour involves seeking out, choosing, obtaining and preparing food, followed by the initiation of a meal. Behaviours that accompany the actual eating process are associated with the desire to eat and a feeling of fullness (choice, deciding on the size of a meal, ceasing intake). The feeling of fullness corresponds to the dynamic aspect of the state of satiety; it is characterised by a slower rate of ingestion and fewer mouthfuls in a given unit of time. Lastly, post-ingestive behaviour is associated with a feeling of satiety characterised by the absence of the desire to eat and a state of well-being and somnolence.

In each of these three phases of the ingestion process, appetite is regulated and a feeling of satiety conditioned on an ongoing basis. This is illustrated by the association between flavour and calorie density. In nature, sweet and fatty foods normally presuppose a certain calorie value, whereas salty foods presuppose mineral content [22]. Nevertheless, the relation between sensorial properties and the calorie or nutritive value of food is not so straightforward. It involves implicit, Pavlov-esque conditioning that takes place gradually throughout eating experiences and by way of post-ingestive metabolic effects. That is how experts have been able to shed light on how people acquire preferences and are conditioned to feel sated. Palatability is defined as the pleasure procured from the sensorial perception of food in standardised conditions. Research has shown that palatability is not solely a response to an internal physiological state (nutritional or energy needs), but part of a reward system for sensorial stimulations. The pleasure derived from eating can be seen as the interaction between desire and appreciation. If one of the two is lacking, there is no reward [23].

As a general rule, the greater the palatability of a food, the greater the quantity consumed. Since foods that are sweet, fatty or salty tend to be much more palatable to consumers, sometimes to the point of creating addictions in the case of sweet and fatty foods, it is paradoxical that the higher the calorie content of a food, the greater the quantity consumed! In order to tackle obesity, exposure of the general population to sweet and fatty foods must be kept in check. This in turn would allow people to shape their preferences for and behaviour toward these kinds of foods.

Eating Habits and the Risk of Overnutrition

Beyond trends in macronutrients and large food groups, people have developed new eating habits which heighten the risk of overnutrition. Namely, this involves snacking, a habit often linked to eating outside of the home, ever-larger portions of food and the introduction of "new" foods. Snacking refers to eating more or less regularly between meals, a habit on the rise. The impact of snacking is difficult to assess. Nevertheless, some experts point to the fact that snacks are higher in calories and do not lend well to a balanced diet. Moreover, since snacking involves the intake of food in limited quantities, the feeling of satiety is never reached. Other studies show, however, that this is compensated by lighter main meals. Portion size, at the distribution and consumption stage, has a positive impact on the quantities of food actually ingested, as shown by the case of French fries [24].

Lastly, the introduction of "new" foods on the market, or the disproportionate rise in the consumption of certain nutrients, has caught the attention of scientists and public authorities concerned about public health. Sweetened beverages are a case in point, the consumption of which has been on the rise in many countries for the past few decades. In France, children (3–17 years of age) currently consume between 200 and 250 ml of sweetened drinks per day, accounting for slightly more than 20 % of their total liquid intake. These sweetened beverages are full of simple sugars (saccharose, fructose, glucose syrup, glucose-fructose syrup), and lead to a significant increase in overall calorie intake.

The Effects of Environment on Eating Habits

The presence of food stimuli in an individual's surrounding environment is an obvious factor in determining eating habits. Several other factors, which may have little to do with food or nutrition, may also affect food-related choices, e.g. frequency and size of meals [25]. The proliferation of easily accessible food sources is likely to stimulate consumption, especially between meals, and encourage people to choose foods that are easy to consume and require little preparation [26]. These food products, often dense in calories, are consumed quickly, anywhere and at any time [27].

Several studies have proven that the availability of easily accessible foods alone stimulates consumption. What is not known, however, is whether this occasional rise in consumption is compensated elsewhere, or whether it affects an individual's calorie count in the long term. In a recent study, Story et al. [28] suggested that actions taken by public authorities to improve a population's food habits should also aim to create an environment conducive to healthy eating. Another documented example is that of the quality of attention paid to the act of eating: when distracted, people tend to eat more. Television is a key factor here.

Marketing and Food Consumption Behaviours

The approach developed by marketing experts is more comprehensive and inductive than that of traditional economics. By looking at the hypotheses and results of research carried out by economists, sociologists, psychologists and nutritionists, they focus on deciphering the logic behind the series of small decisions taken by consumers before and during the act of purchase, while managing food stocks, and at the moment of consumption. Particular attention is paid to consumer perception, the influence of environment, and consumer preferences that are in a constant state of flux. The normative implications of marketing research are clear: the lack of control consumers have on their food choices warrants intervention on the part of public authorities in order to protect consumers from themselves as well as from those who supply food.

This conclusion is supported by the most recent theoretical and empirical developments in economics, which call into question the traditional notion of "consumer as king". In both cases, the consumer spends time and money in order to satisfy personal preferences in terms of pleasure, health and nutrition. The comparison between cost and benefit governs choices, which, in turn, influence future preferences. Economic studies are often based on the assumption that when it comes to pleasure, preferences are stable and unchanged by the context in which purchasing and consumption take place. Put simply, consumer tastes do not change. This hypothesis is too categorical. In reality, preferences are formed gradually by consumer experiences, generating habits that are relatively stable to be sure, but also malleable. Food marketing experts know how to exploit the associations consumers make between, on the one hand, the form, aroma and colour of foods, and, on the other hand, the pleasure of eating—associations that can trigger impulse buying or unplanned consumption.

Along similar lines, it would be logical to assume that proven risks related to overnutrition are a strong incentive to change behaviours. The notion of risk on the part of consumers is nevertheless highly complex, and how consumers deal with those food safety risks is subjective and personal. It depends on how likely the consumer perceives these risks and their potential consequences to be (generally underestimated), as well as the value placed on staying healthy in the long term—a factor that varies widely. In addition to sensorial and nutritional preferences and the amount of information available, two other factors come into play in consumer choices: income and price. How do budgetary and price constraints affect consumer choices? Research provides some very clear answers: consumers are quick to react to price, and when the cost of a food product rises, consumption drops. Lastly, in societies where salaried employment is the norm, time spent on food and eating is restricted. Nowadays, it is less worthwhile for people to fix meals at home than to buy industrially produced meals near the workplace. Although food preparation still occupies first place among all domestic activities (approximately 50 min per day in France), it is definitely losing ground. What is more, accumulated gains in productivity in the food industry have led to a rapid drop in processed food prices. This is not the case for unrefined food products in their natural state. This trend of relative prices promotes the consumption of prepared foods, eating outside of the home and snacking.

Processed Foods and Calorie Excess

How are food processing technologies and consumer behaviour related? It is a known fact that the form and texture of food have an influence on how much of it is ingested. In particular, foods in liquid form are consumed in larger quantities, leading to greater calorie intake. The consumption of meals of equal calorie value but presented in different forms and textures lead to significant physiological variations, especially oxidative metabolism. The easier a food is to consume, the greater the quantity ingested, and therefore the greater the potential calorie intake. The structure of a food also has an impact on the availability of macronutrients and a feeling of fullness and satiety. Nowadays, it is widely accepted that foods with a low glycemic index curb hunger and are therefore ingested in lesser quantities.

Food structure has an impact on both the assimilation and digestibility of lipids and proteins, although how it affects fullness and satiety has yet to be explored in any depth. Fibre also comes into play, procuring a greater feeling of satiety while diminishing hunger and calorie intake. When it comes to additives, the use of flavouring, whether artificial or natural, even in small quantities, acts on the olfactory system to boost the acceptability of a food. The quest for maximal palatability can therefore contribute to detrimental habits such as overeating. To sum up, it seems that food technologies, especially during the secondary processing stage, have an impact on consumer food habits and the assimilation of nutrients via several channels, without the consumers being entirely aware of this. These mechanisms, because they are so complex, warrant further study to better understand the subtle interaction between texture and flavour and how they influence food consumption. While it is regrettable that industrially processed food products too often contribute to an increase in calorie intake, this also represents a major driver of change and an opportunity to improve the quality of food supply. This is key to changing food and eating habits for the better in terms of nutritional value.

Similar trends are being seen with other determining factors of food consumption behaviours. Taken together, they upset diets and lead to excessive calorie intake. The matter is exacerbated by two major and well-established trends: increasingly sedentary lifestyles and pervasive consumerism. Awareness of these health hazards is growing and more and more people are paying attention to their weight, taking physical exercise and trying to improve their lifestyle. However, there is still a lack of documented statistics of the impact of this. Moreover, whole segments of the population are pro-active about finding new ways to obtain food (i.e. opting for local food products), and there is renewed interest in the nutritional quality of food.

Overnutrition and Health

Overweight and Obesity

Overweight and obesity result from an imbalance between calorie intake and expended energy. Because each situation is so different, with a variety of factors coming into play, it is not always possible to determine to what extent increasingly sedentary lifestyles and overnutrition affect matters. The blame is often put on factors such as increased use of motorised vehicles, urbanisation and a trend toward sedentary work and leisure activities. Other studies point the finger solely at increased intake to explain overweight and obesity. In reality, both trends must be taken into consideration together. According to the World Health Organisation (WHO), in 2005 there were 1.6 billion overweight adults in the world (body mass index or BMI[2] above 25), of which 400 million were obese (BMI above 30). Forecasts sounded the alarm bell, with an expected 2.3 billion overweight people and 700 million obese people by 2015. In France, the prevalence of reported cases of obesity rose from 8.5 % in 1997 to 14.5 % in 2009[3]. Nevertheless, France still has one of the lowest obesity rates in Europe (see Table 2.1).

Until the end of the 1960s, obesity among the adult population in the US was stable, at around 15 %, rising to 35 % in the early years of the new millenium [29]. Since then, it seems to have levelled off [30], proving the most pessimistic of forecasts wrong. This recent stabilisation must not, however, lead to an underestimation of the health risks that may well lie ahead for the US population. Indeed, there is a lag in the time between the onset of obesity and its consequences on health (metabolic, vascular, osteo-articular and neoplastic), indicating that a worsening of certain health indices can be expected [31]. Models have shown that the positive health effects linked to a decrease in certain risk factors, first and foremost smoking, are largely offset by overweight and obesity [32]. In the US, over two-thirds of the adult population is currently overweight (obese and non-obese people alike).

In the past, overweight and obesity were considered the hallmark of wealthy nations. Today, these health risks are on the rise at an alarming rate in developing

[2] BMI is calculated by an individual's body weight (in kilograms) divided by the square of his/her height (in metres).

[3] Figures from the national survey ObEpi-Roche on the prevalence of overweight and obesity in France.

Table 2.1 Obesity and food consumption in the world. (Source: OECD, FAO and other data, compiled in Etil, 2010, Food Consumption and Health. In J. Lusk, J. Roosen and J. Shogren (editors), Oxford Handbook of the Economics of Food and Agricultural Policy, Oxford Univ. Press)

Country	Obese adults (prevalence in %)		Overweight non-obese adults (prevalence in %)		Obese and/ or overweight children (7-11 years) (prevalence in %)	Calorie supply in kcal/capita/ day (% from fats)	
Year	1980	2006	1980	2006	2000	1980	2005
Australia	8.3	18.7	28.0	34.4	26.2	3,051 (33%)	3,084 (39.1%)
Canada	13.8	23.1	35.4	36.1	25.1	2,946 (37%)	3,552 (37.1%)
France	6.5	11.5	26.9	31.5	19.0	3,376 (39%)	3,603 (40.7%)
Germany	na	13.6	na	36.0	16.0	3,338 (37%)	3,510 (35.9%)
Greece	na	16.4	na	41.3	31.0	3,216 (35%)	3,700 (35.7%)
Italy	7.1	10.2	27.4	35.0	36.0	3,589 (32%)	3,685 (38.4%)
Japan	2.0	3.4	15.6	21.8	17.8	2,720 (23%)	2,743 (28.0%)
The Netherlands	5.1	11.3	28.2	35.2	12.0	3,071 (38%)	3,240 (38.2%)
Norway	na	9.0	na	34.0	18.5	3,350 (40%)	3,478 (37.4%)
United Kingdom	7.0	24.0	29.0	38.0	20.0	3,159 (39%)	3,421 (36.5%)
United States	15.0	34.3	32.4	33.0	15.2	3,155 (36%)	3,855 (39.4%)

na not available

countries, especially in urban settings where the food transitions discussed earlier are gaining pace. Emerging economies are the first to feel the effects of this, but poor nations are not spared either. According to the OECD, seven Mexicans out of ten are overweight. The figure stands at 50% in Brazil, Russia and South Africa, approximately 28% in China and slightly less than 15% in India. In South America, North Africa, India, China and elsewhere, with the exception of sub-Sahelian Africa, malnutrition due to overeating is more common than undernutrition. Mexico, the country in the world where overweight and obesity are most prevalent after the US, ranks first for childhood obesity, and diabetes is now the second cause of mortality after hypertension. In China, 200 million people are believed to be overweight, of which 90 million are obese, an increase of 39% and 97% respectively since 1992[4].

Overweight, Obesity and Related Illnesses

Research on eating habits and health indicators poses a methodological problem since correlations between variables do not always imply a causal relationship. Analyses are based on a multitude of factors, both diet- and health-related. When an entire population is taken into consideration, the annual incidence of most illnesses is relatively low, calling accuracy into question when it comes to determining

[4] Le Monde 12 July 2010.

the relationship between diet and health. Moreover, a number of factors influence health indicators, for example improved screening practices, disease prevention or therapeutic innovations, all of which make it more difficult to determine the role diet plays.

The sheer diversity of eating habits, the inherent difficulty in assessing them, and ever-changing trends—not to mention how those factors affect health across an entire population—make conducting research on food-related behaviour a genuine challenge. To remedy this, mass numbers of people must be monitored and their eating habits well documented, ensuring that analyses are accurate and reliable. Nevertheless, the overall findings could not be clearer: the food transitions underway in wealthy and poor nations alike are detrimental to health. More specifically, excessive calorie intake has a definite influence on a number of degenerative diseases such as type-2 diabetes, cardio-vascular disease, certain types of cancer and articular disease.

Diabetes

Obesity results from an excess of fat cells in the body which produce free fatty acids. These cells hinder the performance of insulin, the hormone that keeps blood sugar levels in check, gradually compromising its effectiveness. Uncontrolled sugar levels lead to the onset of type-2 diabetes, also known as sugar diabetes or insulin-resistant diabetes, which accounts for 90 % of all diabetes cases today. Epidemic diabetes and obesity follow the same course, which has led to the coining of a new term: "diabesity". Determining a causal relationship is, however, a more complex matter. Other variables come into play, such as old age, sedentary lifestyles, stress, genetic predisposition, etc. Nevertheless, diabetes and overweight do have this in common: they both share two of the four causes of insulin resistance which are eating disorders and a lack of physical exercise.

Cardiovascular Disease

Resistance to insulin is also at the root of many cardiovascular complications associated with obesity. It can lead to the production of a particular type of lipoprotein in the liver that produces "bad" cholesterol in the form of fatty deposits, called atheroma, in the arteries. Excessive amounts of "bad" cholesterol is a proven high-risk factor of cardio-vascular disease. In addition, obesity also triggers hypertension, which in turn exacerbates cardio-vascular risk.

Cancer

Numerous studies have confirmed the link between excessive weight and cancer risk, but once again it is difficult to assess the degree to which overweight and obesity play a role since many other factors come into play, the first among which are a lack of physical exercise and poor eating habits. It is more accurate to speak of associations, and experts concur: there is indeed an association between the frequency of some types of cancer (oesophagus, thyroid, colon, rectum, pancreas, etc.) and overweight and obesity [33]. For every 5 kg/m^2 of BMI, the risk of cancer of the oesophagus increases by 52 % for men and 51 % for women. Likewise, the risk of thyroid cancer increases 33 % for men and 14 % for women.

The "double burden" of Disease in Developing Countries

Developing countries are witnessing a surge in chronic diseases linked to excessive calorie intake, overweight and obesity. Practically speaking, many low- and middle-income countries are currently struggling with what WHO calls a "double burden" of disease, the prevalence of which is defined as the number of persons afflicted by a given disease at a given time. While undernutrition, malnutrition and infectious diseases continue to take their toll, risk factors of noncommunicable diseases increase under the combined influence of factors such as overeating or imbalanced diets, overweight or obesity, a lack of physical activity, and more generally a shift toward westernized habits and lifestyles, especially in urban settings. Increasingly, situations of undernutrition/malnutrition and overnutrition/overweight exist side by side within the same country, community, or even household. It is not rare to find homes where a pregnant woman or infant suffers from undernutrition (inadequate nutrition to satisfy energy requirements) while an older child's diet is rich in high-sugar, high-fat, calorie-dense foods that are cheap but poor in micronutrients. The prevalence of childhood and adolescent obesity is on the rise while malnutrition persists (relative to undernutrition, malnutrition refers to calorie intake that is not only insufficient but unsuited to needs). Indeed, developing countries are facing a triple challenge: accelerated food transitions (excessive calorie intake relative to energy requirements), nutritional imbalances (imbalanced or deficient diets) and of course undernutrition. The issue of hunger and solutions for tackling hunger is the subject of Chapter 9. The next paragraphs explore how to deal with the problem of overnutrition.

Taking Action: A How To

The consequences of excessive calorie intake are extremely grave and constitute a major public health challenge, for developed and developing countries alike. The urgency of the situation is only heightened by the fact that these trends are usually fast-paced. The determining factors responsible for these imbalances have already been outlined, but they may also be seen as windows of opportunity to take action, starting with the consumer, who is ultimately key when it comes to food and nutrition. At this level, matters are anything but simple; the many small decisions consumers make on a daily basis are subject to context, environment, modes of distribution and the constraints and vagaries of everyday life.

Those on the supply side of agricultural products, be they farmers or industrial players, distributors or restaurateurs, must also be taken into consideration. They certainly have a major role in implementing important changes, but are they really able or willing to do so? Is bringing about change in their best interest? Lastly, there are public authorities, whose duty it is to come forward in the name of public health and use all the educational and regulatory tools and incentives to face the issue and stamp out the epidemic. These three categories do not have equal clout in all countries. In-depth studies carried out in industrialised nations serve as a solid basis for action, as they provide thorough knowledge of target systems and populations. The same level of knowledge is unfortunately not available in many developing countries. The actions outlined below are based primarily, but not exclusively, on experiences in France.

Driving Change at Consumer Level

Understanding Trends in the Consumer/Food Relationship

A series of recent health crises in the European Union (mad cow disease in 1996, foot-and-mouth disease in 2001, bird flu in late 2005, etc.) coupled with a surge in overweight, obesity and related degenerative diseases, have resulted in new public awareness of the key role that diet plays in health. This awareness, although perhaps late in coming, presumably marks the beginning of a new era with modified behaviours on the part of the consumer-citizen and decidedly pro-active food policies. After its first National Nutrition and Health Programme (PNNS) in 2001, France, torn between its culinary tradition and a growing prevalence of food-related illnesses, decided to send out a loud and clear message reminiscent of British and American societies, who have entirely different food cultures. When it comes to food, the French place great emphasis on the importance of tradition, sharing and pleasure. A medical discourse that champions health at the expense of cultural or social aspects or pleasure comes at the risk of undermining the very food culture of the French.

The impact of the messages about health sent out by French authorities, but also by industrial players who have taken a stance on the issue, is revealed by people's interpretation of the open question: *what does it mean to "eat well"?* This question has been asked since 1998 in surveys on food-related behaviours in France carried out by Credoc, the French Research Institute for the Study and Monitoring of Living Standards [34]. Until 2003, the vast majority of answers emphasised the importance of shared pleasure. Things changed in 2007, when people started to speak of "balance" in their diets. Today, greater focus is placed on functional foods and the limitations they imply as compared to the 1990s, notably among youth. This reflects, in part, a heightened awareness when it comes to diet and nutrition but may also be explained as a generational shift in mental representations: less emphasis was placed on health in the 1990s, and the issue regained importance in the first decade of the new millennium.

Not all consumers are alike. Likewise, there are differences between perceptions and actual behaviour. Lahlou [35] has observed that "housewives with little free time use processed foods (especially frozen foods), even though they claim not to like them." Likewise, Raude and Denizeau [36] have shown that those consumers who claim to avoid delicatessen meats for health reasons are not those who eat them the least. It is precisely because people feel guilty, or have a negative opinion, about certain nutritional practices, that it is unlikely that trends in France will follow the American model. Nevertheless, achieving a better understanding of how the relationship between consumers and food evolves remains a challenge. In general, according to surveys conducted among the French on eating habits, a growing number of people believe their diets to be "poor".

A Clearer Message About the Risk of Overnutrition

A heated public debate about health and nutrition has been taking place in developed countries on a non-stop basis. Governments, consumerist movements, physicians, industrial players, distributors, the media and advertising are constantly contributing to it, often sending out mixed messages to the public. Contradictory information runs rife in the press, not only between different sources and different articles, but often within the very same article: sources are not always cited and accuracy is compromised by omissions, over-simplification and sweeping generalisations. Providers of information (journalists) and those on the receiving end (readers) act as both agents and factors of these distortions. Consumers are thus faced with risks that can be classified according to two main criteria: a health risk *versus* a nutritional risk and a long-term *versus* a short-term risk. Food poisoning is a short-term health risk while obesity is a long-term nutritional risk. How they cope with these risks is a fundamentally subjective matter, depending not only on the value individuals place on staying healthy but also on the perceived likelihood of risks and potential consequences. Evaluations are based on the incomplete or misleading information at their disposal.

Three lessons may be drawn from studies carried out on consumer behaviour in the face of risk: consumers are uncomfortable with risk that is beyond their control; consumers underestimate the risks they run by their own habits; consumers overestimate rare and unfamiliar risks and underestimate common and familiar risks. These three perspectives explain in part the relationship between consumers and food-related technological innovations, especially when these innovations blur the line between what happens on the farm and what ends up on their forks. To the consumer mind, technological innovation and loss of control mean higher risk. From this perspective, during a health crisis, perceived risk is much greater than actual risk for some people. The risk factor gets blown out of proportion when assessing the quality of a food product. Lastly, this explains why outbreaks of new health risks are more likely to influence consumer food habits more than awareness campaigns that serve as reminders of well-known, familiar risks, e.g. alcohol abuse or a diet that is high in fat, sugar or salt. While no one can deny the importance and necessity of clear, accurate and accessible information about risk, the previous paragraphs reveal that this is not enough to reach targets and significantly change consumer behaviour, especially if information is limited to nutritional aspects alone.

Tapping into the Benefits of Nutrition Education

Nutrition education, a form of intervention widely used by public health professionals, has been the subject of many studies. It entails working in groups within schools, neighbourhoods or specific communities to inform and educate people, and enhance pre-existing favourable conditions in order to improve behaviours. In the majority of cases, the impact of this intervention is gauged in terms of knowledge gained, attitudes and behaviours. Studies show that while nutrition education is indeed effective in boosting knowledge, it is less so in changing actual behaviours. Based on a meta-analysis of articles and reports on nutritional programmes implemented in the UK since 2001, it has been concluded that this lack of effectiveness is linked to the failure to recognize the determining factors that structure food choices, and to social and emotional factors as well [37].

Moreover, nutrition education places nutrition and health at the very core of the question, the underlying principle being that reason guides people to choose healthy foods. This approach may not be appropriate for children, whose reasoning when it comes to health differs from that of adults, and for whom health and taste are mutually exclusive [38]. Nutrition education which takes cultural differences into account can be relatively effective, as the Irish programme *Nutrition Education at Primary School* (NEAPS) has shown: some 453 children aged 8–10 were given 20 lessons over a period of 10 weeks; 3 months after the end of the programme, they were found to eat more fruit and vegetables and fewer salty snacks.

Sensory Education: A Driver of Change, Especially Among Children

Sensory education is well rooted in history. It is based on the intuitive hypothesis that food-related behaviour goes beyond nutritional aspects and involves sensory aspects as well. The role of food is not simply to satisfy nutritional requirements but bring pleasure. What are the sensory properties of food that give pleasure to children? Although it is tempting to think, and even to conclude, that an innate predilection for calorie-dense foods high in sugar and fat govern children's food preferences and choices, this is not at all the case. Numerous studies have shown, in fact, that the food preferences of babies and children can and do change with time. The same holds true for adults. The challenge of sensory training is clear: steer children's dietary preferences toward healthier foods. This idea is not new. Over 30 years ago in France, Jacques Puisais developed a classification method geared toward children aged 7–11 to differentiate between tastes. The goal of the method was to stimulate children's curiosity about food.

These "taste classes" draw children's attention. Significant progress has been reported in the vast majority of cases: improved verbal expression of sensations, increased willingness to try new foods and an interest in helping to prepare meals. In 2004, two major projects were launched almost simultaneously, one in Finland and the other in France under the aegis of INRA (the EduSens project). Both initiatives were geared toward children aged 8–12, and used a longitudinal approach over several years to determine whether or not observed effects, if any were recorded, stood the test of time. In the Finnish project, five lessons served as a complement to sensory education, itself administered in two phases. Each lesson aimed to educate and inform about a certain type of food, how it is made, how to describe its taste and how it can be modified by different procedures. The French project consisted of three independent experiments based on the "taste classes" method and two derived methods, "taste cafeteria" and "taste families". Both projects confirmed that sensory education, especially coupled with other information on food behaviours, has a definite impact, notably heightened interest in a wide variety of foods. Despite these positive results, it would be a far stretch to conclude that sensory education alone can bring about a lasting improvement in eating habits. Nevertheless, the method can and must serve as a stepping stone to change.

Social Marketing: Recommendations for Taking Action

Social marketing goes back to the 1970s, when it was widely used in British and American societies as an effective teaching tool for changing behaviours and modifying social norms [39]. When it comes to putting theory into practice, however, the method has its limits, especially with regard to eating habits. In order for this theoretical approach and framework for action to be effective, it must go beyond the strategic analysis of data and identification of obstacles and levers of behavioural change. To bring about lasting change in individual and collective habits within a social group, or in social norms, things must be taken a step further—social marketing must be combined with a multilevel strategy founded on the active participation of all players, with goals for the short, medium and long terms. Action must be taken where it counts: at the heart of families in their immediate settings [40].

In terms of eating habits and public health, social marketing is more than just a concept, it is a philosophy of change that seeks to promote behaviour that is conducive to good health. It is also about fostering a desire for that change. To this end, it proposes a five-step strategy—the "5 Ps"—built around a people-centred frame of thinking and a thorough understanding of personal motivation and the cultural aspects that influence day-to-day behaviour [41]. Each of the 5 Ps (product, place, price, promotion, and partnership) affords a different point of view, builds upon and enhances understanding, and contributes to the ultimate goal of defining the marketing mix. This mix includes targets, the strategic platform that serves as the basis for intervention, and the corresponding operational system that optimises impact on individual and collective behaviours. According to some experts, such as Cairns and Stead [42], social marketing is an effective technique for altering behaviour and limiting overweight. This perspective is supported by the results of an experiment carried out in the US, *Towards a Healthy Diet* [43], which focused primarily on restaurants, schools and health services. It sought to promote culinary practices that take the nutritional value of foods into account, and coupled this with nutrition education and community events. Evaluation based on a sample population of 1,137 individuals focused primarily on opinions, behaviours and attitudes, and showed that more people frequented "healthy" restaurants and canteens, and their intake of fruit increased.

Several similar interventions were carried out in the US between 1997 and 1999 among 1,109 students between the ages of 11 and 14 in some 24 schools. Low-fat foods were supplied and sold at all meals and anywhere food was made available (i.e. cafeterias, adjoining shops, etc.). The interventions were designed to get both students and personnel involved, and included a physical component as well. Information on nutrition, however, was not provided. The interventions had a positive impact on the physical activity of students and a negative impact on the BMI of boys. There was no significant effect on the consumption of high-fat foods, but additional research is needed to understand why [44]. Social marketing and multi-level strategies are but tools that alone cannot change behaviours and eating habits or alter lifestyles in any significant or lasting way. They must be part of a broader effort designed to get all players involved and make a contribution to change [45].

Product Labels: A Valuable Source of Information

Consumers who pay attention to the link between diet and health, and to the related risks they run with their own behaviour, are naturally more receptive to practical information on the consequences of their food choices and the nutritional breakdown of food. It comes as no surprise, then, that people who attribute greater value to their health and body, i.e. women, well-educated consumers, and/or those with higher incomes, read nutrition labels and are generally more informed about nutrition. How nutrition labels affect food choices is not so easy to determine, especially since consumers tend to evaluate products in an oversimplified, categorical manner, deeming them good or bad, healthy or unhealthy [46]. Information about nutrition and public health campaigns is often in competition with messages sent out by companies via advertising and clever packaging designed to attract the consumer. This competition fosters what is known as the "halo effect", which refers to the name of a food product, or unproven claims about the product, holding greater sway on consumer expectations in terms of pleasure or nutritional value than a detailed breakdown of nutritional content. Unfortunately, these expectations are unfounded. Faced with complex nutrition labelling, consumers tend to focus on one indicator, for example added preservatives or fat content (a particularly important criterion in determining what is healthy versus what is not healthy [47]). Given this, it is hardly surprising that consumers assume a product is high in nutritional quality based on isolated indicators [48]. The calorie content of foods deemed healthy is therefore systematically underestimated, and that of foods deemed unhealthy overestimated [49, 50].

The argument for standardising nutrition labelling is very convincing: the market only functions if consumers are well informed. However, nutrition labels have little bearing on the nutritional quality of people's food choices when they are not used or misused, which happens to be the case. This is most likely due to information overload: messages about health and prevention must vie with advertising for consumers' attention. Likewise, nutrition labels, which are often overly complex, are in competition with unproven claims about a product's nutritional quality. While nutrition labelling is mandatory in the US, it is optional in Europe. Nevertheless, in France, close to 80 % of products produced by different sectors examined by the French Observatory of Food Quality (OQALI) have nutrition labels. In addition, more and more companies are committed to keeping consumers informed via nutrition labels. Nonetheless, they are more common with national brand products (and mid-range and high-end products) than with discount or generic products.

More generally speaking, public campaigns to build awareness about nutrition can upset balances in the market for products that are recommended, or, on the contrary, deemed a risk to public health. One of the first consequences of these campaigns is an increasingly fragmented market, linked to a small, privileged segment of society—the richest and/or most informed—willing and able to pay for products believed to be beneficial to health. This increase in demand pushes the prices of recommended products up, in turn making them less accessible to those segments of the population that are least receptive to campaigns. As a result, the overall average quality of a population's diet can be seen to improve while that of the underprivileged categories of society witnesses a deterioration, be it absolute or relative [51].

The effects described above depend largely on the industrial structure of the sector and the level of competition that prevails. Research carried out in the US on the effects of campaigns promoting whole grain products (bread, flour, pasta, rice, etc.) have shown that they lead to modifications in existing products and the introduction of new products without increasing prices. Companies have been stepping up competition by promoting whole grain products. In a sector where the elasticity of supply is high, competition for gaining greater market share has no effect on price [52].

Targeting Industrial Players and Food Availability

It is unrealistic to assume that consumer preferences cannot change. Sensory perceptions, which directly influence the quantity of food consumed, are inextricably linked to context. It follows from this that they are also malleable. What is more, consumers learn to associate certain sensory properties with certain foods over the course of time. Past experiences shape present tastes, generating relatively stable eating patterns which are reinforced by social norms. Some studies point to the fact that certain foods that are particularly palatable, especially those high in fat or sugar, can create addictions. Food marketing experts know how to exploit the associations consumers make between the form, aroma and colour of foods, and the pleasure in partaking of them. Seeing or smelling food alone can trigger impulse buying or unplanned consumption [53, 54]. These stimuli amplify feelings of hunger by setting off dopamine circuits associated with pleasure and reward, especially in people whose calorie intake is restricted [55, 56]. If these interactions can lead to poor eating habits, there is no reason why they cannot be manipulated and used as leverage to do the opposite and bring about positive change.

Engage Industry in Improving the Quality of Food Supply

When it comes to regulating food supply, the relationship between public authorities and companies can take several forms. While some plead for corporate self-regulation, others are in favour of imposing compulsory government regulations. Still others are for coordinated action between the public and private sectors [57]. The argument for corporate self-regulation is part of the broader issue of Corporate Social Responsibility, which aims to reduce the risk of controversy in the public eye, protect the reputation of brands, promote a positive image among shareholders and employees, and lastly to avoid the application of more stringent regulations in the future [58, 59]. Reconciling self-regulation with the laws of the market and competition is only possible, however, if solutions are found that allow companies to prosper and consumers to have a greater say in their food choices, and therefore health. In such a win-win situation, public intervention becomes a moot point [60]. This seems feasible provided at least a portion of the consumer population is able to pay more for better quality foods in terms of nutrition, and costs do not get pushed up significantly.

Those who argue for public regulatory intervention point to the lack of effort on the part of companies in areas other than the ones listed above, e.g. advertising which targets children, or the incompatibility, inherent to a free market, between health-related expectations and company objectives [61–64]. There is no lack of examples to support their claims. In the US, companies were quick to respond to the ban on trans fatty acids in several states with technical solutions ready at hand. Cutting the sugar or salt content of products is a more complicated matter, not in terms of implementation but gaining acceptance. Indeed, consumers are slow to accept such changes since they tend to blunt sensory pleasure, leaving companies little leeway for modifying products. Between these two extremes, a system of joint regulation between public and private players via partnerships leads to a two-fold awareness. First, companies become aware of their responsibility to contribute to improving the health of populations, or at least to not compromising it. Second, public authorities become aware that companies are in the best position to assess which actions work best in light of the economic constraints of the market.

Several initiatives to promote joint public/private regulation have been taken, either as part of voluntary projects or by way of nutrition charters. Voluntary projects aim to improve the nutritional quality of products (reduced salt or fatty acid content, etc.). They are public and unenforced. Commitment charters are designed to influence impact, measured by indicators correlated to a population's state of health; it is up to companies to take action and punitive measures are taken if there is a breach of discipline. Inter-company agreements to boost minimum quality standards are difficult to implement since the temptation is great for companies to go along with them without truly making a commitment. One way to remedy this is to monitor and publish the efforts made by individual players in different sectors, for example via an observatory for food quality, thus setting an example for others to follow.

In France, 15 commitment charters were signed between February 2007 and February 2010. Depending on the sector, they represent between 5 and 15 % of the market share of related products. These commitments, fully implemented, are far from insignificant: they involve taking 10,000 t of sugar, 400 t of fatty substances and 200 t of sodium off the market. Naturally, the more companies and sectors involved, the more effective these commitment charters are in driving change. In light of this, it is not the number of individual charters signed by players in the food industry that counts, but collective agreements designed to improve the nutritional quality of the most poorly positioned products within each sector. Generic products are particularly problematic since they do not fall within the range of products that come with nutritional claims and/or that are designed with the well-informed consumer in mind. This involves a commercial risk for industrial players in that the modification of a product can influence consumer taste perceptions and alter shelf life. In such a situation, public intervention must have a regulatory component, e.g. imposing minimum quality standards, and incentives to improve, on a voluntary basis, the quality of products.

Regulating the Environment Where Consumer Choices are Made

Given that consumer choices are influenced by systematic biases and that their preferences change according to the environment in which those biases develop, it is perfectly legitimate to question how food marketing should be regulated. This is all the more true given the fact, discussed earlier, that people's knowledge about nutrition is limited. Two types of intervention in particular have been examined: on the one hand, regulations, or even a ban, on advertising that promotes high-fat and/or high-sugar foods, and, on the other hand, regulations on the physical environment where choices are made, i.e. the way in which different options are presented to the consumer in stores [61, 65]. A large portion of advertisements geared toward children and adolescents promotes foods that are high in fat, added sugar or salt; the figure in France stands at approximately 85 %. The problem is exacerbated by other factors: these advertisements are usually broadcast outside of normal mealtimes and they reinforce the association between the promoted product and the benefits they allegedly procure, such as relaxation, happiness, an athletic image and/or good health [65]. Although a causal relationship with consumer decisions is indeed difficult to establish, several experimental studies confirm the theory that such a relationship does exist between advertising and consumer eating habits. In light of this, it is logical to look to regulating advertising in order to limit negative behaviours.

The second course of action involves the physical environment where consumers make choices. Seeing and smelling food greatly increases the chances of it being consumed. It follows that greater visibility of healthier alternatives also increases the chances of consumers taking notice of them. In large and very large supermarkets, fruits and vegetables should therefore be put on more prominent display, and snack foods less so. When it comes to restaurants, one possibility is to keep the number of fast food establishments near schools in check. This measure was proven to work in California: young people with a fast food restaurant within a 0.5 mile radius of their school were found to eat less fruit and vegetables, drink more soft drinks, and have a greater tendency to be overweight or obese. A policy to limit the proximity between schools and fast food establishments would therefore lead to a drop in adolescent overweight and obesity. The need for such a policy is even more pressing in disadvantaged areas where fast food restaurants abound; it is precisely in these neighbourhoods that obesity is most prevalent [66, 67]. The study of systematic biases that influence individual consumer choices provides an irrefutable normative basis for nutritional policies because it challenges the theory of "consumer as king". Such studies show that actions should be geared more to the environment of consumer choices than to the market itself. This type of intervention would lead to fewer distortions than those that focus on price—traditionally the preferred approach of economists. That said, care must be taken to ensure that regulations in the name of public health do not have the opposite effect of that intended: leaving consumers prey to the manipulation of distributors and severely curtailing their freedom to make their own choices!

Product Availability at the Consumption Stage

Places where people gather are ideal ground for promoting healthy food. This is especially true for schools, where several initiatives to promote fruits and vegetables have been taken. In a recent meta-analysis, 10 out of 15 cases showed an increase in the number of fruit and vegetable portions consumed: between one and three extra servings per day. Results were even more promising when a multilevel approach was taken over a substantial amount of time, combining several initiatives such as promoting fresh fruit and water and discouraging sweetened beverages [68]. Moreover, follow-up studies showed that this change for the better continued 2 years after the interventions. Efficiency was further improved when initiatives were combined with organisational changes in schools, instruction on diet and nutrition, the adoption of official policies and partnerships with local food and nutrition services [69].

In the same vein, automatic vending machines in schools have gained a bad reputation for promoting calorie-dense foods. In France, vending machines that sell fruit have met with some success, but current legislation prohibits their installation in schools. Within the framework of pro-active, voluntary initiatives, making fruits and vegetables regularly available in schools has had success when it comes to fruit, less so with vegetables (based on experiences in Denmark and Norway). Likewise, several studies show that the promotion of fruit and vegetables where adults eat brings about the desired result and initiatives taken in the workplace or on university campuses are more effective than those taken in shops [70]. Lastly, smaller portions and packaging containers in shops, vending machines and dining establishments have also been seen to have positive effects, since people tend to eat more when portions and containers are large [71].

Tapping into the Power of Price

This is a prime area of economic research which, in a broad sense, focuses on food consumption behaviours and the policies that are likely to influence them. Studies that examine the consequences of public intervention on consumer demand often take it for granted that supply remains unchanged, in terms of price or otherwise. This assumption is not always well founded, and in some cases the reaction of companies must also be taken into consideration. In other words, the impact of measures taken must be assessed by looking at the big picture: the interplay between supply and demand. Several recent studies have illustrated the dynamics of supply and demand and the importance of balance when it comes to price, taking all adjustments into account.

Public food policy oriented toward supply via prices aims to change consumer habits and improve health, including for those who are not well informed or aware of the link between diet and health [72]. This supply-oriented approach is an example of "passive prevention", which can be beneficial to consumers but does not require any significant effort on their part. The reduction of health inequalities is often cited as justification for such an approach. Nevertheless, the debate about public intervention at supply level is still going strong [60, 73] and examples at international level are few and far between. There are two non-exclusive ways to manipulate the price of products: subsidies and taxes. Several studies have been carried out on food stamp programmes, which are, practically speaking, subsidies to income. Two principle lessons have been drawn from studies carried out in the US: the health benefits of food stamp programmes among qualifying populations are offset, in part, by a negative carryover effect among middle-income populations who are not eligible for such programmes. In other words, a rise in demand for recommended products among consumers who benefit from food stamps drives up prices [74]. A similar risk, albeit theoretical, was found in France in the case of fruits and vegetables. However, it has yet to be empirically confirmed (or disproved) [51].

Few studies have been carried out on the potential effects of nutritional taxes on corporate behaviour. To be sure, this is an area that merits further research. Recent work on this topic by S. Sugarman illustrates the relevance of this point [75]. This US scholar, focusing on one particular market leader in a concentrated industry, showed that companies, whose primary concern is to maintain demand, will not always raise the price of products to compensate for higher taxes. In fact they are more likely to look for ways to cut production costs by compromising the quality of products, thus offsetting the beneficial effect of the tax on consumer health. This theory has yet to be supported by concrete figures.

Bolstering Impact with a Multilevel Approach: Easier Said than Done

One thing is clear from this analysis of different strategies to improve the impact of diet on health: each one is highly unlikely to make a difference if implemented on its own. Clearly, then, a multilevel approach combining several strategies at once to tackle the issues of overnutrition and related illnesses would be much more effective. While no one would deny that such an approach makes sense, establishing a causal relationship between multilevel programmes and public health is another, much more complicated, matter. Some are convinced that a multilevel approach is what is needed [76], especially when it comes to children. Others have doubts about the causal relationship between childhood obesity and environment [77].

Today, with the benefit of hindsight, we are able to assess the impact of wide-scale projects combining several actions implemented in eating environments outside of the home, both in schools and in the workplace. As far as schools go, several programmes deployed in the US have met with success by combining nutrition education, access to "healthy" products, marketing adapted to children and modifying the school environment [78] or promoting physical exercise, nutrition education, programmes to educate and inform parents and providing advice on diet and nutrition [79]. In Belgium, the aggregated school socioeconomic status programme (AGG SES), which examined the influence of availability of food items, nutrition education and adolescent food habits, has also proven somewhat effective in the case of secondary school students. With primary school children, however, its impact was not significant [80]. In Scotland, a multilevel approach geared toward children (aged 6–7 and 10–11), their parents and teachers led to a significant increase in fruit consumption [81].

When it comes to combined programmes in the workplace, they are found to be more effective when people are given an active role in implementing them (increased involvement) and when their scope is broadened to include surrounding social environments (families, neighbours, etc.). Organisational barriers on site must also be overcome [82]. Combined programmes which aim to boost the consumption of fruit and vegetables have also met with success in Finland, as long as they took into account the socioeconomic aspects and cultural differences of targeted populations [83]. The diversity of situations goes to show that there is no one-size-fits-all solution. This has been further confirmed by the teachings of the EU programme Identification and prevention of dietary and lifestyle induced health effects in children and infants (IDEFICS), with the participation of eight member states (Germany, Belgium, Cyprus, Spain, Estonia, Hungary, Italy and Sweden). It is based on the analysis of 20 groups of children aged 2–8, and 36 groups of parents. The programme sheds light on the different obstacles to improving food habits: lack of time, financial constraints, personal preferences, influence of family and peers, school policies, etc. [84].

The challenge before us is daunting: provide food for 9 billion people. The stakes are enormous, both in the quantitative and qualitative sense. That is why we must look for ways to combat behaviour that leads to malnutrition issues rooted, ironically, in the overabundance that today is the cause of a growing number of public health issues all over the world. The issue is all the more pressing considering that a rise in overweight and obesity is occurring side by side with undernutrition, which affects close to 1 billion people, primarily in poor countries but also among the disadvantaged classes of wealthy nations. These two issues will more than likely have to be addressed simultaneously in order to boost the effectiveness of proposed solutions and achieve real results.

Several lessons from this chapter must be kept in mind when drawing up a plan of action. First and foremost is the fact that the consumer is not king, in that he has little control over his food choices. Consumers are subject to a wide range of constraints, both conscious and unconscious, and are influenced by education, information, advertising and the actions of those who supply food. Direct intervention on food prices is one possible solution, but it must be well thought out in order to limit indirect adverse effects, and implemented in tandem with other actions that target the entire spectrum of variables that influence consumer behaviour and environment: nutrition policies unrelated to price, campaigns that educate and inform about health, easier access to "healthy" food and more difficult access to "unhealthy" food, regulations for advertising and packaging, reduced food portions, etc.

Taking action strictly geared toward changing people's behaviour is not enough; action must also be geared toward environment. The multilevel programmes discussed above will be all the more effective if they are implemented within a framework that gets all players involved, is long-term, and adapted to specific communities in terms of the people themselves but also of their environments. Lastly, public intervention must also have a regulatory component to incite all players—producers, agribusinesses and distributors—to step up and take responsibility. Despite auspicious initiatives that have, for the most part, met with success, such programmes are still all too rare. The gap must be bridged between being aware of overnutrition and actually doing something about it. Although there is still much to learn and uncertainties abound, this must not serve as an excuse to postpone—or worse, renounce—tackling the issue.

References

1. Combris P (2006) Le poids des contraintes économiques dans les choix alimentaires. Cahiers de Nutrition et de Diététique 41(5):279–284
2. Balanza R, Garcia-Lorda P, Pérez-Rodrigo C, Aranceta J, Bonet MB, Salas-Salvadó J (2007) Trends in food availability determined by the Food and Agriculture Organization's food balance sheets in Mediterranean Europe in comparison with other European Areas. Pub Health Nutr 10(02):168–176
3. Grignon C, Grignon C (1999) Long-term trends in food consumption: a French portrait. Food Foodways 8(3):151–174
4. Combris P, Amiot-Carlin MJ, Baberger-Gateau P, Bouhsina Z, Caillavet F, Causse M, Chervin C, Chevassus-Lozza E, Codron JM, Dallongeville J, Dauchet L, Delcourt C, Demigné C, Dosba F, Dupont C, Gurviez P, Martel P, Mauget JC, Nicklaus S, Padilla M, Renard C, Requillart V, Roy C, Soler LG, Verger P, Volatier JL, Sabbagh C, Savini I, Fournier D,

Champenois A (2008) Les fruits et légumes dans l'alimentation: enjeux et déterminants de la consommation. Éditions Quae, Versailles, 127 p
5. Besson D (2008) Le repas depuis 45 ans: moins de produits frais, plus de plats préparés. Insee Première (1208):1–4
6. Nichèle V, Andrieu E, Boizot-Szantai C, Caillavet F, Darmon N (2008) L'évolution des achats alimentaires: 30 ans d'enquêtes auprès des ménages en France. Cahiers de nutrition et de diététique 43(3):123–130
7. Herpin N, Verger D (2008) Consommation et modes de vie en France: une approche économique et sociologique sur un demi-siècle. Éditions la Découverte, collection Grands repères, Paris, 256 p
8. Bellamy V, Léveillé L (2007) Enquête budget de famille 2006. Insee Résultat (73).
9. Courson JP (1990) L'hypermarché se rapproche, l'épicier quitte le village. In Données sociales, Insee, Paris, pp 202–205
10. Lemel Y, Mercier MA (1987) L'approvisionnement alimentaire. In Données sociales, Insee, Paris, pp 406–409
11. Lutinier B (2002) Les petites entreprises du commerce depuis 30 ans: beaucoup moins d'épiceries, un peu plus de fleuristes. Insee Première (831):1–4
12. Guilbert PD, Perrin-Escalo HD, Lamoureux PP (2002) Baromètre nutrition santé. Édition 2002, Inpes
13. Escalon H, Bossard C, Beck F, Bachelot-Narquin RP (2009) Baromètre nutrition santé. Édition 2008, Inpes
14. Dubuisson-Quellier S (2009) Circuits courts: partager les responsabilités entre agriculteurs et consommateurs. In Déméter 2009, pp 87–111
15. Herpin N (1988) Le repas comme institution: Compte-rendu d'une enquête exploratoire. Revue française de sociologie 29(3):501–521
16. Chauffaut D (2001) L'évolution des relations entre générations dans un contexte de mutation du cycle de vie. Cahier de recherche. Credoc, Paris, p 163
17. Volatier JL (1999) Le repas traditionnel se porte encore bien. Consommation & modes de vie (132):1–4
18. Popkin BM (2008) Will China's nutrition transition overwhelm its health care system and slow economic growth? Health Aff 27(4):1064–76
19. Kearney J (2010) Food consumption and drivers. Philos Trans Royal Soc B 375(1554):2793–2808
20. Cépède M, Lengellé M (1970) L'économie de l'alimentation. Presses universitaires de France, Paris, p 128
21. Périssé J, Sizaret F, François P (1969) Effet du revenu sur la structure de la ration alimentaire, Rome, FAO. Bulletin de nutrition 7(3):1–10
22. Apfelbaum M, Romon M, Dubus M (2004) Diététique et nutrition. Éditions Masson, Abrégés de médecine, 6e édition, p 535
23. Finlayson GKN, Blundell JE (2007) Liking versus wanting: importance for human appetite control and weight regulation. Neurosci Behav Rev 31:987–1002
24. Freedman MR, Brochado C (2010) Reducing portion size reduces food intake and plate waste. Obesity 18(9):1864–1866
25. Kamphuis CBM, Giskes K, de Bruijn G-J, Wendel-Vos W, Brug J, van Lenthe FJ (2006) Environmental determinants of fruit and vegetable consumption among adults: a systematic review. Br J Nutr 96(4):620–635
26. Cohen D, Farley TA 2008 Eating as an automatic behavior. Prev Chron Dis 5(1):A23
27. Popkin BM, Duffey K, Gordon-Larsen P 2005 Environmental influences on food choice, physical activity and energy balance. Physiol Behav 86(5):603–613
28. Story M, Kaphingst KM, Robinson-O'Brien R, Glanz K (2008) Creating healthy food and eating environmental approaches. Annu Rev Pub Health 29:253–272
29. Ogden CL, Carroll MD, Curtin LR, McDowell MA, Tabak CJ, Flegal KM (2006) Prevalence of overweight and obesity in the United States, 1999–2004. J Am Med Assoc 295(13):1549–1555

30. Flegal KM, Carroll MD, Ogden CL, Curtin LR (2010) Prevalence and trends in obesity among US adults, 1999–2008. J Am Med Assoc 303(3):235–241
31. Van Dam RM, Willett WC (2009) Unmet potential for cardiovascular disease prevention in the United States. Circulation 120(13):1171–1173
32. Stewart ST, Cutler DM, Rosen AB (2009) Forecasting the effects of obesity and smoking on US life expectancy. N Engl J Med 361(23):2252–2260
33. Reneham A et al (2008) Body-mass index and incidence of cancer: a systematic review and meta-analysis of prospective observational studies. Lancet 371:569–578
34. Mathé T, Pilorin T, Hébel P, Denizeau M (2008) Du discours nutritionnel aux représentations de l'alimentation. Cahier de recherche C252, Credoc, Paris
35. Lahlou S (1992) Si/alors: bien manger? Application d'une nouvelle méthode d'analyse des représentations sociales à un corpus constitué des associations libres de 2 000 individus. Cahier de recherche C34, Credoc, Paris
36. Raude J, Denizeau M (2008) La perception des risques alimentaires: une influence complexe et incertaine sur le comportement des consommateurs. Économies et sociétés, systèmes agroalimentaires (AG) 30:1271–2188
37. Attree P (2006) A critical analysis of UK public health policies in relation to diet and nutrition in low-income households. Matern Child Nutr 2(2):67–78
38. Wardle J, Huon G (2000) An experimental investigation of the influence of health information on children's taste preferences. Health Educ Res 15(1):39–44
39. Kotler P, Armstrong G (2005) Principles of marketing. Prentice Hall International, 11th edn, p 768
40. Henley N, Raffin S (Forthcoming) EPODE (Ensemble, prévenons l'obésité des enfants or "Together, Let's Prevent Childhood Obesity") Case Study: Preventing Childhood Obesity. J Mark Manag
41. Glanz K, Rimer B, Lewis F (2002) Health behavior and health education. theory, research and practice. Wiley, San Fransisco
42. Cairns G, Stead M (2009) Obesity and social marketing: works in progress. Proc Nutr Soc 68:11–16
43. Dunt D, Day N et al (1999) Evaluation of a community-based health promotion program supporting public policy initiatives for a healthy diet. Health Promot Intern 14(4):317–327
44. Sallis JF, McKenzie TL et al (2003) Environmental interventions for eating and physical activity—a randomized controlled trial in middle schools. Am J Prev Med 24(3):209–217
45. Swinburn B, Egger G, Raza F (1999) Dissecting obesogenic environments: the development and application of a framework for identifying and prioritizing environmental interventions for obesity. Prev Med 29(6):563–570
46. Wertenbroch K (1998) Consumption self-control by rationing purchase quantities of virtue and vice. Mark Sci 17(4):317–337
47. Oakes ME, Slotterback CS (2005) Too good to be true: dose insensitivity and stereotypical thinking of foods' capacity to promote weight gain. Food Qual Prefer 16(8):675–681
48. Keller SB, Landry M, Olson J, Velliquette AM, Burton S, Andrews JC (1997) The effects of nutrition package claims, nutrition facts panels, and motivation to process nutrition information on consumer product evaluations. J Pub Policy Mark 16(2):256–279
49. Wansink B, Chandon P (2006) Can 'Low-Fat' nutrition labels lead to obesity? J Mark Res 43(4):605–617
50. Carels RA, Konrad K, Harper J (2007) Individual differences in food perceptions and calorie estimation: an examination of dieting status, weight and gender. Appetite 49(2):450–458
51. Dallongeville J, Dauchet L, de Mouzon O, Requillart V, Soler LG (2010) Increasing fruit and vegetable consumption: a cost-effectiveness analysis of public policies. Eur J Pub Health (published online on February 25, 2010)
52. Mancino L, Kuchler F, Leibtag E (2008) Getting consumers to eat more whole-grains: the role of policy, information, and food manufacturers. Food Policy 33(6):489–496
53. Cornell CE, Rodin J, Weingarten H (1989) Stimulus-induced eating when satiated. Physiol Behav 45(4):695–704

54. Painter JE, Wansink B, Hieggelke JB (2002) How visibility and convenience influence candy consumption. Appetite 38(3):237–238
55. Volkow ND, Wise RA (2005) How can drug addiction help us understand obesity? Nat Neurosci 8(5):555–560
56. Polivy J, Herman CP, Coelho JS (2008) Caloric restriction in the presence of attractive food cues: external cues, eating, and weight. Physiol Behav 94(5):729–733
57. Lang T (2006) Food, the law and public health: three models of the relationship. Pub Health 120(Suppl 1):30–40
58. Herrick C (2009) Shifting blame/selling health: corporate social responsibility in the age of obesity. Sociol Health Fit 31(1):51–65
59. Mikkelsen BE (2005) Declining role of governments in promoting healthy eating: time to rethink the role of the food industry? Scandinavian J Nutr 49(3):127–130
60. Pramming S (2008) Should we use regulation to demand improved public health outcomes from industry? No. BMJ 337:a1761
61. Harris JL, Pomeranz JL, Lobstein T, Brownell KD (2009) A crisis in the marketplace: how food marketing contributes to childhood obesity and what can be done. Annu Rev Pub Health 30:211–225
62. Hawkes C (2007) Regulating and litigating in the public interest: regulating food marketing to young people worldwide: trends and policy drivers. Am J Pub Health 97(11):1962–1973
63. Pomeranz JL, Brownell KD (2008) Legal and public health considerations affecting the success, reach, and impact of menu-labeling laws. Am J Pub Health 98(9):1578–1583
64. Stanley F, Daube M (2009) Should industry care for children? Public health advocacy and law in Australia. Pub Health 123(3):283–286
65. Seiders K, Petty RD (2004) Obesity and the role of food marketing: a policy analysis of issues and remedies. J Pub Policy Market 23(2):153–169
66. Kumanyika S, Grier S (2006) Targeting interventions for ethnic minority and low-income populations. Future Child 16(1):187–207
67. Davis B, Carpenter C (2009) Proximity of fast-food restaurants to schools and adolescent obesity. Am J Pub Health 99(3):505–510
68. Knai C, Pomerleau J et al (2006) Getting children to eat more fruit and vegetables: a systematic review. Prev Med 42(2):85–95
69. Laurence S, Peterken R et al (2007) Fresh kids: the efficacy of a health promoting schools approach to increasing consumption of fruit and water in Australia. Health Promot Intern 22(3):218–226
70. Seymour JD, Yaroch AL, Serdula M, Blanck HM, Khan LK (2004) Impact of nutrition environmental interventions on point-of-purchase behavior in adults: a review. Prev Med 39(Suppl 2):108–136
71. Devitt AA, Mattes RD (2004) Effects of food unit size and energy density on intake in humans. Appetite 42(2):213–220
72. Barling D, Lang T, Caraher M (2002) Joined-up food policy? The trials of governance, public policy and the food system. Soc Policy Adm 36(6):556–574
73. Sugarman S (2008) Should we demand improved public health outcomes from industry with more regulation? Yes. BMJ 337:a1750
74. Alston JM, Sumner DA, Vosti SA (2008) Farm subsidies and obesity in the United States: National evidence and international comparisons. Food Policy 33(6):470–479
75. Sugarman S (2009) No more business as usual: enticing companies to sharply lower the public health costs of the products they sell. Pub Health 123(3):275–279
76. Cawley J (2006) Markets and childhood obesity policy. Futur Child 16(1):69–88
77. Sallis JF, Glanz K (2006) The role of built environments in physical activity, eating, and obesity in childhood. Futur Child 16(1):89–108
78. Briggs M, Safaii SA et al (2003) Position of the American Dietetic Association, Society for Nutrition Education, and American School Food Service Association; Nutrition services: an essential component of comprehensive school health programs. J Nutr Educ Behav 35(2):57–67

79. Ritchie LD, Crawford PB et al (2006) Position of the American Dietetic Association: Individual-, family-, school-, and community-based interventions for pediatric overweight. J Am Diet Assoc 106(6):925–945
80. Vereecken CA, Bobelijn K, Maes L (2005) School food policy at primary and secondary school in Belgium-Flanders: Does it influence young people's habits? Eur J Clin Nutr 59(2):271–277
81. Anderson AS, Porteous LEG et al (2005) The impact of a school-based nutrition education intervention on dietary intake and cognitive and attitudinal variables relating to fruits and vegetables. Pub Health Nutr 8(6):650–656
82. Sorensen G, Linnan L et al (2004) Worksite-based research and initiatives to increase fruit and vegetable consumption. Prev Med 39:94–100
83. Prattala R (2003) Dietary changes in finland—success stories and future challenges. Appetite 41(3):245–249
84. Haerens L, De Bourdeaudhuij I et al (2009) Developing the IDEFICS community-based intervention program to enhance eating behaviors in 2- to 8-year-old children: findings from focus groups with children and parents. Health Educ Res 24(3):381–393

Chapter 3
Reducing Losses and Waste at Consumption, Distribution and Processing Levels

The sustainable scenario put forth by the foresight study Agrimonde (Agrimonde 1 scenario) is based on the bold assumption that all countries in the world will have a food availability of 3,000 kcal per day per person by 2050—a ration the FAO deems sufficient. This presupposes an increase in developing countries, notably sub-Saharan Africa where daily food availability currently stands below 2,500 kcal, and a decrease in developed countries where it stands at about 4,000 kcal per day. Does this mean that the Agrimonde scenario calls for voluntary cutbacks in the richest nations? Not necessarily. First of all, let us recall that food availability and actual consumption are not the same thing. Food availability refers to the quantity of food available for the entire population of a country, not the quantity of food that is actually consumed. Daily actual consumption in France, for example, stood at 2,200 kcal per adult in 2007 according to the Inca2 study [1]. The Agrimonde 1 scenario goes a step further and suggests that a cut back in losses and waste across the entire food supply chain, from farm to fork, would lead to more satisfying actual consumption derived from 3,000 kcal of availability. This chapter explores the sources of these losses and waste, the reasons behind them, and possible solutions to remedy the problem.

"Losses and waste" refer to both the losses that occur independently of human activity (natural disasters, health and hygiene crises, defects in industrial procedures, etc.), and to those caused by poor management or neglect. In all cases, the food lost could have been used to nourish people, which is cause for concern considering that undernutrition persists and the global population is growing. Moreover, these losses and waste take a toll on the resources used to produce food: water, energy, soil, farm inputs, labour, etc. In the US, for example, the energy required to produce food accounts for 10% of total energy consumption [2]. These losses and waste are also a source of pollution: greenhouse gas emissions, the use of chemical inputs, etc. Lastly, they constitute a significant economic burden when it comes to treating and disposing of the waste—an issue that is particularly problematic at the beginning of the twenty-first century and subject to ever-more stringent regulations. Analysis reveals that in some cases there may be underlying cultural factors and a

M. Guillou, G. Matheron, *The World's Challenge*,
DOI 10.1007/978-94-017-8569-3_3, © Éditions Quæ, 2014

certain reasoning behind such waste: having "too much" to eat is a sign of wealth and power, heavily-stocked food aisles is a sign of abundance that reassures consumers, destroying surplus harvest keeps prices in check, etc.

The Scope of the Issue

In the early 2000s, only 2,000 of the 4,600 kcal theoretically available per day per person worldwide were actually "used" (see Chap. 1). This difference can be explained by three factors: primary losses incurred after harvest (600 kcal); resources used to feed animals, who produce less kilocalories in the form of animal products than they consume; and secondary losses at distribution and consumption level (800 kcal). While primary losses are most common in developing countries—a topic explored in Chap. 4—secondary losses are more the norm in industrialised nations. These figures represent a global average; data vary slightly when the focus is exclusively on secondary losses in rich nations. For example, the Economic Research Service (ERS) of the United States Department of Agriculture (USDA) estimates that 27 % of edible foods went to waste in 1995 due to secondary losses at distribution level, in the restaurant and catering industry and in homes [3]. Cuellar et al. [2] went on to calculate that this percentage represents 2 % of total energy consumption in the US, a figure that is far from insignificant and moreover probably an underestimate since it is based on outdated (1995) and incomplete data on losses and waste. As a comparison, this quantity of wasted energy exceeds the amount of energy produced in the US by corn-based bioethanol or offshore drilling.

Another very recent study (2009) carried out in the US was based not on a direct calculation of what is thrown away and wasted, but on estimates obtained by comparing the amount of available food in the US (data based on FAO statistics) to the amount of food actually consumed by the population. This amount is evaluated using a mathematical model for metabolism, which compares the quantity of ingested food with body weight. Based on this, Hall et al. [4] estimate that in 1974, approximately 900 kcal per day per person went to waste in the US. Thirty years later, in 2003, the figure rose to 1,400 kcal per day per person, i.e. 40 % of food availability and a much higher percentage than that calculated by Kantor et al. [3].

Calculating losses and waste at household level is difficult and costly, which explains why this kind of study is rare. Let us look at the findings of the UK-based Waste and Resource Action Programme (WRAP). The rubbish bins of some 2,138 households were collected and analysed in 2007. The results showed that of the 21.7 million t of food purchased each year in the UK, households threw away 6.7 million t (31 %). These figures, however, must be put into perspective by taking note of the fact that what is thrown away is not always edible. Of the 6.7 million t of rubbish, only 4.1 million t were perfectly edible, 1.3 million t potentially edible (bread crust, potato skins, etc.)—bearing in mind that what is potentially edible is relative—and 1.3 million t inedible (inedible peelings, bones, tea bags, etc.). The annual cost of these losses and waste is considerable, approximately 10 billion pounds (11 billion €) [5]. Expressed in kilocalories, the quantity of edible and potentially edible waste reaches 7,696 billion kcal, which amounts to 357 kcal per day per person in the UK [6]. These WRAP figures are confirmed by those of a study on the contents of household waste in Wallonia (Belgium) in 2003 and 2004: 60.2 kg of kitchen refuse were generated per person per year [7]. In France, a study carried out by the French Environment and Energy Management Agency (Ademe) showed that in 2007–2008, 7 kg of unopened packaged food products were thrown out per person per year [8].

Losses at Different Stages and Technical Solutions Currently Implemented

Losses at Production Level

Agricultural producers in developed countries have a solid handle on harvesting, storage and transport techniques. Nevertheless, significant losses have been observed at this stage, mostly for economic reasons.

Withdrawing Fruit and Vegetables to Regulate the Market

The fruit and vegetable industry is subject, in France as in most countries, to strong annual fluctuations both in terms of production and sales and marketing. This has led to the implementation of structural and economic measures to regulate the quantity of produce introduced on the market. Structural measures include two methods in particular: staggered harvests and diversified processing and conservation techniques (fruit juice, canned sauces, etc.).

In the case of overproduction, economic measures consist primarily of financing the withdrawal of produce from the market to protect prices. During the 2006–2007 agricultural campaign, financial compensation was provided for more than 16,000 t of withdrawn fruit and vegetables in France, principally nectarines, cauliflower, peaches, clementines and apples [9]. Several things can happen to the withdrawn produce: it can be donated to charitable associations or non-governmental organisations to benefit those in need in France or in other countries (6.6 % of 16,207 t withdrawn in 2006–2007, mostly cauliflower and apples), or it can be used as animal feed (7.5 %, again cauliflower and apples). Most of it is composted or biodegraded. Between 1986 and 1995, annual production of apples in the Provence-Alpes-Côte d'Azur region of France varied between 430,000 and 600,000 t, with up to 230,000 t disposed of in landfills or the controlled spreading of sludge over fields or fallow land [10]. The new EU common market organisation (CMO), in force since 1 January 2008 in the fruit and vegetable industry, aims to phase out the financing of withdrawn produce from the market. Regulations stipulate that withdrawn produce be either donated to charitable organisations, distilled, used as animal feed or for purposes unrelated to food, or destroyed in an environmentally friendly manner.

Standards

Standards applied in wealthy nations require that fruit and vegetables that do not meet certain criteria in terms of size, colour or shape be withdrawn from traditional marketing networks. This is called sorting or grading. The Amou agricultural cooperative in the Landes region of France estimates that 25–30 % of fruit that it handles is withdrawn from networks during this process [11]. In the case of greenhouse tomatoes, approximately 3 % of production is deemed non-marketable due to serious defects, and 5 % is withdrawn after sorting. Nevertheless, a second option consists in selling the withdrawn produce through local channels. In the case of salad greens, losses in the field vary between 5 % for greenhouse produce and 15 % for open air produce, and sorting and retailer send-backs account for a further 5 %. In order to limit the waste of edible foods and offer consumers a wider choice, EU standards were relaxed on 1 July 2009 for 26 types of fruits and vegetables. As the then European Commissioner for Agriculture, Mariann Fischer Boel, said: "*It makes no sense to throw perfectly good products away, just because they are the 'wrong' shape*". Nevertheless, marketing standards were maintained for other fruits and vegetables, which account for 75 % of production in the European Union (apples, citrus fruits, kiwi, lettuce, peaches, nectarines, pears, strawberries, peppers, grapes and tomatoes). The fruits and vegetables that do not meet standards can, however, be sold in retail networks as long as they are labelled to distinguish them from top-range produce.

Losses at Industrial Food Processing Level

The Challenge of Managing Perishable Foodstuffs

Managing perishable foodstuffs is particularly difficult in a context where the supply of agricultural produce is subject to unpredictable events and fluctuating demand depending on a host of factors (climate, economy, relative prices, etc.). This can sometimes lead to shortages or, inversely, a surplus of stock. The more perishable a food item is, the more problematic the issue. Fruits and vegetables, fresh dairy products and meat pose the greatest challenge.

Hygiene and Food Safety

Food processing operators are responsible for the hygiene and safety of their food products. In the European Union, they are subject to regulations set down by a food "Hygiene Package" adopted on 1 January 2006. Consisting of several bodies of EU legislation, the Hygiene Package is designed to put in place a single and transparent policy on hygiene for human and animal food, and implement an efficient alert system covering the entire food supply chain from start to finish. As a precautionary measure, processors leave very little leeway when establishing a "best before" date for food items, using a "worst case scenario" approach such as leaving groceries in a car parked in the sun for hours on end. In practice if not in theory, food items can be consumed several days after their best-by date if they are properly transported, stocked and prepared. Distributors and consumers therefore end up throwing perfectly edible food away. As explained in Chap. 2, this can be explained by a certain risk-aversion in a context where health and safety objectives and the goal to reduce losses and waste are at odds with one another.

Ready-to-eat Food Products

Today's consumer is on the lookout for food products that are practical, ready-to-cook or ready-to-eat. This leads to an increased volume of raw commodities excluded from the final product: fish filets rather than the whole fish, ready-to-use packaged salad greens rather than a head of lettuce, etc. In France, whole chickens only account for 38 % of poultry sales, whereas chicken pieces and processed chicken products account for 40 % and 20 % respectively [12].

Turning Waste into Co-products or By-products

The manufacture of food items generates a whole host of organic products that can be marketed as food for human or animal consumption [13]. For example, molasses is derived from the processing of sugar beets and contains 50 % sugar (saccharose). Thanks to its chemical and physical makeup, molasses can be marketed in a number of ways: for animal feed, as fermentation substrate for the production of alcohol, glutamate, citric acid, amino acids, baker's yeast and vitamins, but also in non-food industries for water treatment or soil detoxification. The milling industry produces the bran that is used for bran bread. Distilleries that produce *eaux de vie* from pomace provide the grape seeds from which grape seed oil is manufactured. The dairy industry puts many co-products to good use: buttermilk for fermented drinks, cheese parings and fragments for cheese fondues, whey powder as an additive in several food products (preparations for meat products, sausages, soups, etc.) as well as in baby and diet foods. The meat industry also generates a number of co-products that are used in many ways, notably for livestock feed. This list suffices to show that commercial food processors get optimal use out of co-products and by-products, at least in industrialised countries.

Packaging

The disadvantage of food packaging is that it generates waste in the form of paper, cardboard, and plastic. Packaging accounts for approximately 40 % of waste from the restaurant and catering industry. Moreover, whereas paper and cardboard can be recycled, plastic packaging, which is often composite (trays, film, etc.) cannot. On the other hand, packaging allows for easier transport and improved food preservation of products, thereby curbing losses and waste at these two stages. This is especially true for fresh produce. Thus a new dilemma arises: cut losses and food waste or reduce the waste that food packaging generates. Packaging is not the exclusive hallmark of the modern societies of developed nations; its known benefits have a long history all over the world. In tropical regions, plant leaves are widely used to wrap food. In addition to providing a physical barrier, the leaves impart aromas, colouring and enzymes (such as papain), or agents which delay bacterial growth (e.g. some essential oils) to the food contained within them. This vegetal packaging interacts with the food, enhancing its colour and texture, and limiting the development of certain harmful micro-organisms [14].

"Modern" packaging is multi-functional: it protects the product against external aggressions, is user-friendly (convenient serving size, reusable containers), improves preservation under controlled conditions (active packaging), and offers a support to provide consumers with information. New plastic films developed for wrapping fresh products such as meat in controlled and/or modified conditions extend shelf life to beyond 10 days and satisfy criteria imposed by distributors who require a minimum shelf life of one week. Losses, which with traditional packaging can reach 5–10 % in the poultry aisle, are thus kept in check [15].

Losses at Distribution Level

Retailers must also face the difficulty of adapting supply to demand. When it comes to fresh produce, this is particularly challenging. Large quantities of fresh produce are thrown away every day, for reasons that are hard to avoid (rotten fruit, torn packaging), or due to poor management of products that could otherwise have been consumed. Storage and management constraints are such that when a new shipment of fresh produce arrives, it replaces a former batch which is often thrown away. Retailers must also follow through with the marketing strategies they employ: shelves must be well stocked to reassure consumers, and the quantity of goods displayed therefore exceeds demand. Products whose best-by date is drawing near are very often removed from shelves even though they are still perfectly edible, once again to reassure consumers. Nevertheless, supermarkets and hypermarkets are getting into the habit of promoting products nearing expiration at the cost of more hired help.

What proportions has this issue taken on? In the US, the quantities of food waste are far from insignificant (11 % for fresh fruit, 10 % for fresh vegetables, 5 % for meat and seafood), with strong variations between products (0.6 % for sweet corn but over 63 % for mustard greens) [16]. Cutting back on quantities lost is therefore no easy task! As already mentioned, packaging in plastic containers is proven to protect products well, especially for delicate fruit such as blueberries or tomatoes. New varieties with thicker skin that offer better protection and therefore boost shelf life have been developed. The problem is that these advantages often come at the cost of taste. For example, tomatoes modified with a gene to slow ripening are often mealy and less aromatic [17]. It all comes down to a choice between taste and preservation! A short selling season, such as for sweet corn, which is sold from mid-summer to the first frosts, helps to reduce waste because consumers know the product will not be available for long. When it comes to lettuce, limiting shelf room helps limit losses as well, by displaying the product for a shorter time and increasing turnover. This comes at the expense, perhaps, of a feeling of abundance that consumers look for and retailers seek to elicit.

Consumer knowledge also comes into play. Losses for products such as papaya are significant (55 %) since not many people are familiar with this exotic fruit. For meat, losses are greater for veal (25 %) and lamb (12 %) because consumers do not know how to prepare those meats well, and therefore buy less; nevertheless, they are still to be found in food aisles, as part of an effort to broaden the range of products on offer. Inversely, less and less seafood goes to waste—although losses remain more or less substantial, around 9 %—as more seafood dishes enter mainstream cooking. A large quantity of edible food finds its way into the rubbish bins of retailers, restaurants, cafeterias and markets, where they are sometimes retrieved by poor people or militants. Freeganism, a movement born in the US that has slowly spread to other parts of the world (in France, the concept is known as *glanage* and its adherents *glaneurs*), speaks out against waste, and more generally the absurdity of the consumerist model of so-called developed societies, by proving that it is possible to live off the contents of rubbish bins. Driven by ethical, social and economic ideals and drawing upon several different philosophies (anticapitalism, antiglobalization, anarchism, etc.), freeganism makes the bold claim that "the solution to world hunger is in the garbage cans of New York City".

This slogan, true or not, stands in sharp contrast with what appears to be a certain narrow-mindedness of some businesses that have resorted to dousing their rubbish bins with bleach or keeping them under lock and key! Others take a more positive approach by giving away discarded food items that are still edible to charitable organisations for distribution to the needy. Although this may seem like a simple solution, it is not: storage, handling, transport and respecting the cold chain all come at an additional cost. Questions of responsibility and hygiene issues can also arise if distribution deadlines are too long or if people get food poisoning by consuming expired products. This would warrant intervention on the part of public authorities, in the form of regulatory measures or incentives, to encourage businesses to act responsibly when it makes edible waste available to the public.

Losses at the Food Catering Level (Restaurants and Cafeterias)

At this level as well, very stringent hygiene measures implemented to protect consumer health can lead to substantial food waste. Prepared foods spoil easily. Regulations have been set down regarding maximum temperatures for conserving food: in France, +4 °C for mixed salads, +8°C for deserts that contain milk, +63 °C for prepared dishes that are kept hot and − 18 °C for ice cream.[1] All uneaten food exposed

[1] Decrees of 9 May 1995 and 21 December 2009.

to temperatures that do not fall within these ranges, for example in a buffet, must be thrown out. Habit and cultural practices often lead to waste since traditionally hosts take pride in offering their guests lavish meals that far exceed need.

Once prepared, dishes do not keep long. When establishments provide more food than is required, or that is in excess of a day's demand, they have no choice but to throw away the leftovers. In France, a meal served in a restaurant, fast-food establishment, cafeteria or hospital generates 70–330 g of food waste in addition to other types of waste such as packaging, residue from grease traps, etc. Over a one-year period, this represents over 1 million t of waste, of which the majority is thrown away with household waste and unrecovered, for lack of specific regulations and reluctance on the part of restaurant owners and local authorities to make a firm commitment to reducing waste [18]. Several countries have assessed the amount of waste produced in school cafeterias. In Sweden, studies carried out in two school cafeterias and two restaurants showed that only 80 % of food is consumed. Of the 20 % of food left uneaten, half is left on plates while the other half is thrown away in the kitchen during storage or preparation [19]. A bibliographical study carried out by the USDA showed more moderate figures for the US, with an estimated 12 % of calories served to students wasted [20].

Programmes to educate and inform about sustainable development have been organised in schools. These include efforts to reduce food waste and cooking lessons. Initiatives taken in Sweden have shown promising results, as the following two examples show: in one school, teachers asked students to weigh the leftovers on their plates, displaying the figures for all to see in the dining area. This resulted in a 35 % drop in food waste. In another school, leftovers were used to feed chickens, students taking turns to carry the buckets of scraps to the chicken coops. After 1 year, the number of buckets was down to one per day [19]. Lastly, an experiment carried out in a self-service cafeteria of an American university is worthy of note: students who were unwittingly served smaller portions of French fries not only ate less, but also threw away less at the end of their meal [21]. These examples go to show that simple initiatives can and do lead to substantially less wasted food at the end of a meal.

Losses at Consumer Level

The issue of food waste is not new. According to Tristram Stuart [6], the remains of mammoths discovered in the United States show that only very few bones bear any traces of cutting. This would seem to suggest that 12,000 years ago, man left a good part of his prey to rot. Rather than cutting all the meat into strips for drying and preserving, it would seem that our ancestors preferred to move on to make another kill elsewhere. Half-eaten, abandoned mammoth carcasses foreshadow modern day food waste. Waste at the consumer level may also be explained by the fact that initiatives such as weighing leftovers to reduce waste cannot compete with tradition and culture. In times of festivity, serving large quantities of food is a sign of esteem for guests. In wealthy nations, this habit was reinforced after the Second World War under the two-fold influence of a decrease in relative food prices and the illusion of an abundance of food unprecedented in the history of humanity. In France, the share of expenditure on food in household budgets went from 25 % in 1960 to less than 15 % today.

Beyond these historic, cultural and social aspects, a decidedly more practical issue is at hand—that of the margin of error in interpreting dates indicated on packaging. In France, two different dates appear on packaged foods: a "use by" date for perishable food items (meat, ready made meals) that must be respected for reasons of hygiene and food safety, and a "best before" date for products with a longer shelf life (biscuits, pasta, canned goods). The "best before" date is purely for informational purposes, not to be confused with the "use by" date. There is no need to throw away perfectly edible products whose best-by date is past. At worst, the products lose a little in the way of quality, for example biscuits may be less crunchy. A survey conducted in Wallonia (Belgium) in 2007 clearly illustrates a lack of knowledge among consumers: 47 % are under the false impression that food items with opened packaging can still be consumed until the use-by date, and 33 % think, also incorrectly, that a food item whose use-by date is past can no longer be consumed [22]. To complicate matters, in the UK, in addition to these same two dates used in France, two other dates appear: "display until" and "sell by". They are intended for retailers (for managing stock) and do not concern consumers, who need not take them into account [23].

The consumption of certain parts of certain foods that the UK-based WRAP considers "potentially edible", e.g. potato peelings, crusts of bread, certain leftovers, etc., raises another issue. Removing the skin strips fruits and vegetables of a host of nutritious elements that are found in or just beneath the skin, such as carotenoids, polyphenols, fibre, glucosinolates or ascorbic acid [24]. The vitamin C concentration of apple peels is three times that of the actual flesh. Furthermore, as lifestyles in western societies evolve, people are spending less time cooking and lack the culinary skills required to prepare foods that are less prized, or altogether shunned: offal, stale bread, edible leftovers, etc. Is it really necessary to leave a clean plate after all? Parents would most often respond "yes" while children would almost certainly beg to differ! The contradiction between the goals of reducing losses and waste and limiting calorie intake is all too obvious. The solution is to cut back on serving sizes, a proven way to significantly limit the amount of food left over and thrown away after meals.

Change is in Order

Food items that are still edible are nevertheless removed from the food supply chain at different stages of production, processing, distribution and consumption, both inside and outside the home. The quantities involved are considerable. To be sure, this does not mean that these food items go to complete waste. Deformed carrots, yoghurts 3 days past the expiry date or stale bread are put to good use in animal feed or compost piles. Nonetheless, from an economic, environmental and ethical point of view, a better solution would be to reduce losses and waste in the first place, and when this is not possible or is too difficult—for instance due to regulations or prohibitive costs—take the necessary steps to put edible food waste to good use in "other" forms for human consumption before resorting to animal feed, methanization or composting.

As the quantity of waste generated by so-called modern societies increases, more stringent regulations have been put in place in an effort to reverse the trend. The hierarchy of solutions is clear: first, prevention; second, recovering waste for other uses; third, recycling; and lastly, elimination. In the European Union, a longstanding directive (1999/31/CE) provides for a gradual ban on biodegradable waste in landfills. Duly applied, the quantity of biodegradable waste disposed of in landfills should see a 75 % decrease between 1995 and 2016 [25]. The onus is on individual member States to take the necessary steps to reach that goal. France, for instance, will gradually raise its landfill tax from 15 € per tonne of waste in 2009 to 40 € in 2015. Since food waste is included in this biodegradable quotient, this rise in taxes is meant to incite the producers of said waste to find other ways to put it to use rather than dumping it in landfills. Naturally, this will only happen if the tax ultimately represents a greater financial burden than the three alternatives of prevention, recovering waste for other uses and recycling. What is for certain is that this transfer of costs to players in the food industry is a step in the right direction. The next section goes beyond these initiatives and takes a look at how waste can be put to better use firstly for human consumption but also for animal feed, and as a source of energy.

Increasing Human Use of Agricultural and Agri-food Products

Throughout history, food preservation and processing procedures have boosted the life span of agricultural commodities and the diversity of uses for human consumption. Recently, CIRAD and the equipment manufacturer Auriol teamed up to develop a pilot machine that makes crisps from any fruit or vegetable. Such an innovative tool puts fruits and vegetables cast aside during sorting to good use. As a first step, the machine will be used to make melon crisps in Costa Rica, where 200,000 t of melon are removed after sorting every year [26]. The role of distributors when it comes to optimising the use of agricultural and food products for human consumption cannot be stressed enough. This is particularly pertinent for sales of fresh produce whose use-by date is approaching. Nevertheless, such offers are time-sensitive; if they are implemented too late, consumers may well make a purchase, but proceed to throw out the items at home for a lack of time for consuming them. "Three-for-the-price-of-two" sales should also be avoided, as they encourage overconsumption. As a general rule, care must be taken to coordinate actions at different levels of the food supply chain, from production to consumption, to avoid problems at one stage simply getting transferred to another, sometimes in an exacerbated state.

The Role of Public Policy

In the European Union, the European Programme of Food Aid to the Most Deprived Persons (PEAD) was initiated in 1987 to address social issues and work toward goals related to regulating the market. The programme allowed for a barter-like exchange of public intervention stocks of agricultural commodities (cereals, sugar, butter, milk powder and meat) for more or less elaborate finished products (pasta, desserts containing milk, cream desserts, tuna and rice salad, etc.), which were then delivered to charities for distribution to the poorest. In case public intervention stocks were temporarily out of a raw commodity, member states were given a budget to buy finished products directly on the market. Successive reforms of the Common Agricultural Policy (CAP) and the phasedown of intervention stocks have made the initial system obsolete. In 2009, for instance, only sugar stocks were sufficient for trading, and there has been a shortage of meat stocks since 2004. The system has therefore been changed accordingly. In France, a national food aid programme (PNAA) was set up in 2004 with a budget of up to 7–10 million € per annum for the purchase of animal proteins (meat and fish) as well as fruits and vegetables—foods which are often lacking in the diets of those most in need [27]. Fiscal policies that encourage putting edible food waste to good use (subsidies for good practices) and/or discourage other, less virtuous practices—especially dumping edible food in landfills—constitute another conceivable course of action. Implementing these incentives and/or deterrents, however, is a tremendous challenge due to the wide range of players involved and the difficulty in finding an easy-to-use indicator on which to base subsidies and/or taxes.

Educating, Informing and Raising Awareness

Another approach is to educate and inform people. Several countries have taken initiatives along these lines to raise awareness among the general population of the urgency of the issue, complete with practical tips on how individuals can play their part. For example, in the UK, WRAP launched the "Love food hate waste" campaign [28], with attention-grabbing figures such as 420 pounds (approximately 360 €) for the estimated annual cost of food wasted per household. Other organisations in other countries draw analogies to illustrate how high the stakes are. For example, according to the Brussels Institute for Management of the Environment (IBGE), throwing away a loaf of bread is the equivalent of driving a car for 2.24 km in terms of green house gas emissions. IBGE also gives practical tips to consumers via a dedicated Internet site [29] and a brochure entitled "Food and the environment: 65 environmentally-friendly ways to eat well and stay healthy" [30]. Another example, which may seem silly but is far from it, is an initiative taken in the Belgian city of Mouscron, where two laying hens were offered as an incentive to households to cut down on waste. After initial training, volunteers received the two hens provided they agreed to raise and feed them properly, and not cook them before 2 years' time.

Action Aimed at Food Professionals

In addition to consumers and citizens, action should be taken to educate producers, commercial food processors, distributors and those in the restaurant and catering business. Numerous initiatives are geared specifically to these groups, designed to raise awareness of the issue by revealing the cost incurred by losses and food waste, all the while encouraging players to adopt good practices. The stakes are high, not only for those with a vested interest in the issue but for the general public as well. As the President of the Massachusetts Food Association said: "If all 400 supermarkets in Massachusetts recycled their organics, the industry could realise more than $ 4 million in savings per year." The United States Environmental Protection Agency (EPA) also organises an annual contest for companies within the framework of their WasteWise Program [31]. Companies must furnish a progress report presenting figures for cut-backs in food waste, accumulated savings and initiatives taken to promote the programme among employees, customers and suppliers. The agency provides technical support to companies to assist them in their efforts, and publishes several thematic documents to educate and inform, specifically designed with agri-food industry professionals in mind ("Doing what it takes to be WasteWise") [32].

Using Waste to Feed More Animals

According to the United Nations Environment Programme (UNEP), using organic food waste to feed animals is a solution with considerable untapped potential on a global scale. The first and foremost advantage of this solution would be to free up cultivatable land. UNEP estimates that the calories "lost" each year by feeding cereal grains to livestock could cover the needs of half of humanity, i.e 3.5 billion people! [33] Such a figure has definite shock value, but it does raise serious questions in terms of methodology since it is likely based on overestimated equivalents. UNEP also recommends the use of fish waste to feed fish: the 30 million t of fish used as feed each year in fish farms should be used to feed humans instead. It can be replaced with rejected catches thrown overboard, which represents roughly the same quantity.

Composting or Methanization to Avoid Landfill Waste

Composting and/or methanization is another solution to optimise food waste when using it for human or animal consumption is not an option. In many countries, local authorities are trying to promote these methods, for example by organising selective waste sorting, distributing free biodegradable bags for organics or providing households with garden compost units. In most cases, these initiatives go hand in hand with awareness-raising campaigns in order to boost effectiveness. Packaging can hinder composting and methanization. In several countries and regions, agri-food businesses such as commercial food processors and distributors are required by law to recycle biodegradable waste, but packaging must first be removed from cakes, pizzas, sandwiches, yoghurt and ham products. This can incur significant costs in terms of labour if the process is done manually, and remains a major obstacle even where recycling is not mandatory. Nevertheless, it is possible to reduce costs thanks to innovative technologies and automated unpacking plants.

In Alsace, organic waste is banned from landfills, and the first French factory for separating organic waste from packaging opened in September 2010. After waste is separated from packaging by a crushing process, it is transferred to methanization units. Approximately 8,000 establishments have contributed to an organic waste deposit destined for methanization of an estimated 60,000 t per year [34]. Likewise, several outlets in the Auchan retail chain located in the north of France had 3,100 t of organic waste treated in 2009 in a Belgian methanization plant that was also equipped with an automated unpacking machine. Thanks to this operation, landfilling, at a cost of 120 € per tonne, was avoided. As a last example, taken once again in France, the fast-food chain McDonald's intends to treat its food waste by methanization. It is estimated that the fast-food chain generates 43 t of waste that can be methanized per restaurant per year (29 t of cardboard and other packaging, 7 t of organic food waste from kitchens or dining areas, and 7 t of used cooking oil) [35].

The journey from farm to fork is beset with losses and waste accounting for 30% of global food availability, and up to 50% according to the most pessimistic of forecasts. In developing countries, losses and waste occur primarily in the field and immediately upon exit from the field; Chap. 4 will take an in-depth look at this issue. In developed countries, the most significant losses and waste occur at distribution level and during final consumption. Reducing these losses and waste is top priority. When this is not possible, there is a clear hierarchy of solutions: first, reuse for human consumption; second, use as animal feed; third, recycling for use as an energy source or fertilizer. Landfilling should be a last resort. Regulations and incentives should be in line with this list of priorities, while ensuring that competition between companies is not hindered nor consumer excess encouraged. The argument for competition presupposes the need for an analysis of the costs and benefits of alternatives, from both a medium- and long-term point of view, all the while taking expected advances in science and technology into account. "True prices" should be the guiding principle here, that is, the duty of commercial food processors, distributors and players in the restaurant and catering sector to factor in the cost of landfilling to local communities. In keeping with the tenets of the economic theory of induced technical innovation [36], this transfer of landfill costs to players in the field is meant to act as an incentive to come up with new ways to tackle the issue such as reducing waste, using waste as food, and recycling.

When it comes to consumers, a multilevel approach similar to the one described in Chap. 2 to change excessive food consumption patterns to benefit health, is needed here as well, geared at once toward individuals and their environment: education, training, pedagogy, setting examples, regulatory measures and incentives, sorting waste and recycling, etc. Significant headway has been made in many countries but the road ahead is still long. In light of this, priority must be given to solving the dilemma of protecting consumers while reducing consumer waste: how can we prevent hygiene and safety requirements and expiry dates from creating more waste? How can protective packaging be used to optimise food preservation without generating more waste in cardboard or plastic? This is the challenge researchers and society at large must rise to.

References

1. Étude individuelle nationale des consommations alimentaires 2 (Inca2) 2006–2007, septembre 2009
2. Cuellar AD, Webber ME (2010) Wasted food, wasted energy: the embedded energy in food waste in the United States. Environ Sci Technol 44:6464–6469
3. Kantor LS, Lipton K, Manchester A, Oliveira V (1997) Estimating and addressing America's food losses. Food Rev, January-April: 2–12
4. Hall KD, Guo J, Dore M, Chow CC (2009) The progressive increase of food waste in America and its environmental impact. PLoS one 4(11):e7490
5. WRAP, April 2008 (revised July 2008). The food we waste: food waste report v2. Oxon
6. Stuart T (2009) Waste: uncovering the global food scandal. W.W. Norton and Company, 480 p
7. Research Development and Consulting (RDC) (2005) Étude statistique de la composition des ordures ménagères en Région wallonne. Bruxelles
8. Ademe (2009) Campagne nationale de caractérisation des ordures ménagères: résultats année 2007
9. Viniflhor (2007) Rapport d'activité 2007
10. De Cockborne AM, Bruckler L, Sévenier G, Vallès V (2002) Conséquences de l'épandage de pommes de retrait sur la qualité des eaux de drainage. Étude et gestion des sols 9(1):25–41
11. Boyrie-Fénié B et al (2001) Landes—Encyclopédie Bonneton. Christine Bonneton éditeur, Paris
12. Jez C, Beaumont C, Magdelaine P, Paillard S (eds) (2009) La filière avicole française à l'horizon 2025. Rapport du groupe de travail prospective avicole, Inra et Itavi (France), 88 p
13. Réséda (Réseau des organisations professionnelles et interprofessionnelles pour la sécurité et la qualité des denrées animales) et Ademe 2005. Gisements des coproduits, sous-produits et déchets des industries alimentaires. Édition n°1, 126 p
14. Gontard N (2000) Panorama des emballages alimentaires actifs. In: Gontard N (ed) Les emballages actifs. Tech & Doc Lavoisier, pp 1–29
15. Arvaut G (2010) Films rétractables: protéger et séduire. RIA, La revue de l'industrie agroalimentaire, septembre
16. Buzby JC, Wells HF, Axtman B, Mickey J (2009) Supermarket loss estimates for fresh fruits, vegetables, meat, poultry and seafood and their use in the ERS loss-adjusted food availability data. USDA, ERS, Economic Information Bulletin 44, March, 20 p
17. Causse M (2010) Communication aux carrefours de l'innovation agronomique sur la qualité des fruits et légumes. INRA, Avignon, 14 octobre 2010
18. Ademe (2000) La gestion des déchets de la restauration en France: état des lieux. Ademe éditions, Paris
19. Engström R, Carlsson-Kanyama A (2004) Food losses in food service institutions. Examples from Sweden. Food Pol 29:203–213
20. Buzby JC, Guthrie JF (2002) Plate waste in school nutrition programs: final report to Congress. USDA, ERS, March 2002
21. Freedman MR, Brochado C (2010) Reducing portion size reduces food intake and plate waste. Obesity 18(9):1864–1866

22. Centre de recherche et d'information des organisations de consommateurs (CRIOC) (2008) Le gaspillage alimentaire: mieux le comprendre pour mieux le prévenir. Bruxelles, 20 p
23. Food Standards Agency. Eat well, be well. Foodlabels. Labelling terms. http://www.eatwell.gov.uk/foodlabels/labellingterms/. Accessed 30 Aug 2010
24. Combris P, Amiot-Carlin MJ, Baberger-Gateau P, Bouhsina Z, Caillavet F, Causse M, Chervin C, Chevassus-Lozza E, Codron JM, Dallongeville J, Dauchet L, Delcourt C, Demigné C, Dosba F, Dupont C, Gurviez P, Martel P, Mauget JC, Nicklaus S, Padilla M, Renard C, Requillart V, Roy C, Soler LG, Verger P, Volatier JL, Sabbagh C, Savini I, Fournier D, Champenois A (2008) Les fruits et légumes dans l'alimentation: enjeux et déterminants de la consommation. Éditions Quae, Versailles, 127 p
25. Soulage D (2010) Traitement des ordures ménagères: quels choix après le Grenelle? Sénat, Report n°571 of Daniel Soulage carried out in the name of the steering cimmittee on waste, 183 p
26. http://www.cirad.fr/actualites/toutes-les-actualites/communiques-de-presse/2010/chips-de-fruits-et-legumes
27. MAAP (2010) Réduire l'injustice alimentaire. http://agriculture.gouv.fr/reduire-l-injustice-alimentaire. Accessed 3 Sept 2010
28. http://www.lovefoodhatewaste.com/
29. http://www.bruxellesenvironnement.be/Templates/Particuliers/Niveau2.aspx?id=1816&langtype=2060
30. http://documentation.bruxellesenvironnement.be/documents/65_conseils_alimentation_FR.PDF?langtype=2060
31. United States Environmental Protection Agency. Conserving resources, preventing waste. http://www.epa.gov/epawaste/partnerships/wastewise/index.htm. Accessed 12 April 2011
32. http://www.epa.gov/epawaste/partnerships/wastewise/pubs/foodfact.pdf
33. Nellemann C, MacDevette M, Manders T, Eickhout B, Svihus B, Prins AG, Kaltenborn BP (eds) (2009) The environmental food crisis—the environment's role in averting future food crises. A UNEP rapid response assessment. United Nations Environment Programme, GRID-Arendal, www.grida.no
34. Sita—filiale de Suez Environnement (2010) En France, Valorest, la première unité capable de traiter les biodéchets emballés. http://www.sita.fr/document/?f=identite/fr/dp-bio-decoditionneur.pdf. Accessed 22 Dec 2009
35. http://www.cooperonspourlenvironnement.com/index.php/contributions-mcdonalds/11-methaniser-les-dechets-des-restaurants. Accessed 22 Dec 2009
36. Hayami Y, Ruttan W (1998) Agriculture et développement, une approche internationale. Inra Editions, 600 p (French translation)

Chapter 4
Reducing Post-harvest Losses in Developing Nations

In developing countries, food losses occur primarily upstream of industrial food processing and sales and marketing, between the farm and commercial processors or distributors. These "post-harvest" losses are generally kept to a minimum in developed countries and the relatively high-tech agricultural systems of emerging economies. Nevertheless, they cover a wide range of extremely diverse situations linked to type of agriculture, climate, degree of sophistication and organisation of agri-food processing systems, and of course to the crops produced. This diversity makes quantifying losses and understanding their multiple causes difficult and a one-size-fits-all solution will therefore not do. The organisation of food systems in many developing countries is evolving rapidly under the influence, notably, of increasingly urban lifestyles based on western models: markets and small-scale retailers selling local products are losing ground to large retail chains on the outskirts of cities with increasingly diverse suppliers including at international level. Simultaneously, there is a growing trend in eating outside of the home in urban settings (snack bars, cafeterias and "maquis"—informal, makeshift eating establishments popular in Africa that also serve as a forum for exchange, a place for people to meet, etc.).

In this new context of food distribution, can the nature of losses be expected to follow trends in developed nations where, as discussed in previous chapters, losses are incurred primarily at industrial processing level (limited quantities) and to a much larger extent at distribution and consumer levels? Or, on the contrary, will the proximity of urban agriculture with small-scale farming serve to mitigate and recover secondary losses? The question remains unanswered for lack of studies carried out on losses in the urban distribution and consumption systems of developing countries. This chapter explores the issue of post-harvest losses. Before delving into the heart of the matter, let us take a preliminary look at the situation in developed nations.

M. Guillou, G. Matheron, *The World's Challenge*,
DOI 10.1007/978-94-017-8569-3_4, © Éditions Quæ, 2014

Harvest-Related Losses: A Reality in Developed Nations as Well

In developed countries, where heavy investments are made in agriculture, losses during and after harvest are insignificant where cash crops are concerned. These crops are primarily cereal grains, and harvesting, drying and storage procedures are highly industrialised, optimised and efficient. This is not true, however, across the board notably for forage crops. While grazing is the preferred use of grassland, because the part of the plant eaten by the animals has the highest nutritive value (calorie content, digestibility and protein content), this is not the case for forage stored as hay or silage. These stocks are indispensable for rearing ruminants, since the growth of grassland cover is highly seasonal [1]. With haymaking, depending on the harvesting and drying methods used, losses can reach up to 55 % for hay produced from legume crops dried on the ground in humid climates. The figure can drop to 10–20 % for hay that dries rapidly, with the help of conditioning systems that further speed up the drying process. Losses can even be brought down to 5 % if hay is dried in barns, a method that requires significant investment. Keeping losses in check has the added advantage of improved quality (protein content, digestibility and ingestibility) since losses always take the greatest toll on leaf material. When it comes to silage, losses can stand between 16–20 % depending on the harvest method used and dry matter content. Further losses must be factored in due to inedible and unavailable portions which can reach 10 % in the case of bale wrapping.

Keeping losses to a minimum in harvesting forage crops is therefore a key challenge, not only from an economic standpoint but from an environmental one as well. In Sweden, for instance, research and development in animal husbandry for the past few decades have focused primarily on keeping losses down. In general, priority is given to mechanisation, which must be suited to the size of farms, or, in the case of silage, to preservatives. In light of this, industrial drying methods are a practical solution to the problem, since they reduce losses to zero in the field and maintain the nutritive value of the final product. The drawback, however, is the high cost in terms of energy. Pre-curing is a good solution to cut back on energy costs, but intensity and duration must be controlled to avoid leaf loss during harvesting.

Let us return to the issue of post-harvest losses in developing countries and explore why it warrants attention (Table. 4.1).

Why Post-harvest Losses in Developing Countries Matter

According to M. Grolleaud [4], food loss refers to any change or degradation in the quantity, quality or edibility of a food. In this broad sense, post-harvest losses are losses that occur throughout the entire sequence of food production operations, starting with and including harvesting. These losses can be grouped into two categories: "technical" activities (harvesting, drying, threshing, cleaning, storage, processing)

Table 4.1 Average losses and variations in forage production in France according to mode of harvest, conditions and final product (relative to fresh forage). (Sources: Dulphy and Martin-Rosset (2001), Demarquilly et al. (1998) [2, 3])

Harvest and curing conditions	Losses (%)	Effect on protein content (g/kg DM)	Effect on digestibility (% DM)	Effect on protein content (g/kg DM)
Fan-assisted barn drying with additional heat	10–15	−8	−3.9	−7
Ground drying/good weather conditions/forage grass	10–20	−8	−3.9	−13
Ground drying/inclement weather/forage grass	30–55	−27	−10.1	−20

g/kg DM gramme per kilogram of dry matter

and "economic" activities (transport, sales and marketing, quality control, culinary preparations, etc.). The choice of vocabulary to describe each category leaves something to be desired. The *Natural Resources Institute* (NRI)[1] has proposed an alternative classification by making the distinction between physical losses (damaged product leading to a loss in quantity and/or quality with the result that it sells with difficulty if at all), market opportunity losses (sales below the market price) and external losses which affect society as a whole (for example, impact of losses on food security at local and worldwide levels), including via the effects of measures taken to avoid certain losses (impact of pesticides on the environment and health).

In developed countries and in technically developed agri-food sectors of many developing countries, harvest and storage systems are highly efficient and post-harvest losses are an issue primarily in the case of perishable fresh produce throughout the entire supply chain. In other developing countries, the issue takes on other forms. This was the focus of the 1974 World Food Summit organised by the FAO after a string of poor harvests that spanned several years. The Summit gave rise to an ambitious programme entitled *Prevention of Food Losses* designed to reduce global food losses by half in 10 years' time. Curtailing post-harvest losses was recognised as a vital step in achieving food security. Other initiatives have been launched at international level, such as Gasga (Group for assistance on systems relating to grain

[1] Within the framework of a study commissioned by the World Bank on the effectiveness of methods implemented to reduce post-harvest losses; Hodges R. (2010). Postharvest Losses (PHL) Reduction for Cereal Grain Staples in sub-Saharan Africa. Presentation at the conference Reducing Post-Harvest Losses in Grain Supply Chains in Africa, FAO/World Bank, 18–19 March 2010, Rome.

http://www.fao.org/ag/ags/publications/docs/misc/FAO_WB_ph_web.pdf.

after harvest)[2]. Naturally, measuring post-harvest losses is key, if only for an accurate assessment of food quantities readily available at local, regional, national and global level. This is particularly important in poor countries, who struggle to provide an adequate food supply to their populations and depend on food aid. Local missions charged with evaluating agricultural and food supplies (CFSAM, *Crop and Food Supply Assessment Missions*) carried out by the FAO and the World Food Programme (WFP) seek to analyse local situations within the framework of the FAO's *Global Information and Early Warning System* (GIEWS) [5]. How effective public food security policies and similar initiatives in the private sector are depends on the capacity to accurately assess post-harvest losses and understand the underlying factors and mechanisms at work. The previous chapter dealt with secondary losses during processing, distribution and final consumption; the remainder of this chapter will focus on primary losses (in the field and upon exit from the field) in developing countries, referred to as post-harvest losses.

The Challenge of Quantifying Post-harvest Losses

Post-harvest losses vary widely depending on product, climatic conditions, technical equipment, marketing methods, knowledge and expertise, and choices governed by cultural and social values, etc [6]. This chapter will focus primarily on starchy food crops (cereal grains and tubers) which account for the bulk of calorie intake in developing countries. In the 1980s, several post-harvest loss assessment methods were proposed, focusing primarily on grain storage in rural areas [7, 8]. In the vast majority of cases, the methods were based on research carried out by entomologists who sought to reduce losses caused by pests. In the case of grains, it was shown that loss levels are highly dependent on the nature of the plant material (traditional, improved or hybrid varieties). For example, in Malawi, maize losses by weight can vary from 3 % for traditional varieties to more than 20 % for hybrid varieties at the start of the rainy season. The percentage of infested grains rises considerably thereafter, with a particularly higher peak for hybrid varieties than for local or improved varieties, as shown in the table below. Although hybrid varieties have the significant advantage of yielding greater quantities in the field, the fact that they are prone to pest infestation is just as serious a drawback where neither modern and effective pesticides nor a viable stocking system are available.

In order to get a better handle on managing post-harvest losses of starchy food crops, it is necessary at the very least to distinguish between those crops that keep well (cereal grains) and those that do not (roots and tubers). The following table provides such a comparison. Each type of crop responds to storage differently. As

[2] Gasga (ACIAR—Australia, Cirad—France, CRDI—Canada, FAO, GTZ—Germany, KIT—Netherlands, KSU—USA and NRI—UK) sought to improve the technical support provided to developing nations in the sales and marketing, storage, processing and transport of grains. See: Gasga workshop (24–25/02/1983), Slough (GRB), Eschborn (DEU), GTZ, 1984, 136 pages.

Fig. 4.1 Percentage of maize grains attacked by insects in Malawi throughout a season. (Source: Schulten (1982), reproduced with the permission of F. Bourin Publishers)

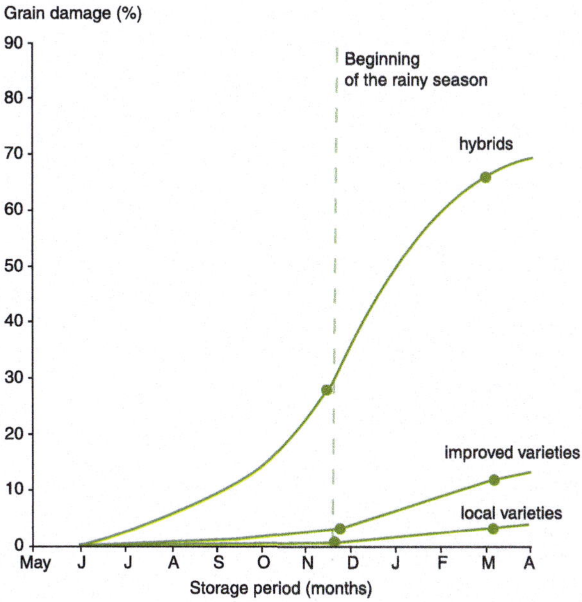

a general rule and for obvious reasons, it is more difficult to store perishable foods, which explains greater losses among this type of crop (Fig. 4.1).

Despite Difficulties in Quantifying Losses, One Thing is Clear: Post-harvest Losses are Significant

It has been shown that the stakes of post-harvest losses are high at the planetary level: approximately 600 kcal per day per person (see Chap. 1). This global perspective masks wide disparities depending on product, place, growing year, etc. As a consequence, extrapolating quantities based on the occasional survey can be misleading if the determining factors of losses are unknown. It is preferable to illustrate how losses can vary depending on the product based on more in-depth surveys and long-term research (Table 4.2).

Cereal Grains

Cereal grains are the primary source of food in developing countries; it comes as no surprise therefore that studies on post-harvest losses have focused more on these crops. In Africa, the Post-Harvest Losses (PHL) [8, 9] network allows for in-depth, long-term studies of losses incurred during storage, since those incurred during other stages (harvest, drying, cleaning and transport) are poorly documented and

Table 4.2 Comparison of storage aptitude of cereal grains (durable food crops) and roots and tubers (perishable food crops). (Source: Grolleaud 1997)

Durable food crops (cereal grains)	Perishable food crops (roots and tubers)
Mainly seasonal harvests, need for prolonged storage	Possibility of permanent or semi-permanent harvests, short-term storage
Preliminary treatment of stored produce rarely necessary (except for threshing)	Conversion into dry products often replaces storage of fresh produce
Fruits with low water content, 10–15 % or less	Fruits with high water content, generally 50–80 %
Small fruits, most often less than 1 g	Voluminous fruits, from 5 g to 5+ kg
Very low respiratory activity of stored fruits, reduced heat	Intensive respiratory activity of stored fruits with heat release, notably in tropical climates
Hard tissue ensuring good protection against bruising	Soft tissue, highly vulnerable
Naturally apt for storage, possibly for several years	Easily perishable produce, naturally apt for storage of several weeks to months depending on species and/or variety
Storage losses generally due to exterior factors (mould, insects and rodents)	Losses due partly to inherent factors (respiration, transpiration, germination), partly to exterior factors (moulds, insects, rodents)

therefore unreliable. In practice, the network furnishes estimates of losses at different stages with a breakdown of post-harvest losses, as illustrated in Table 4.3. At the end of the 9-month storage period, cumulated losses for maize stood at 20 %, of which 4–5 % occurred during storage, and 12 % for sorghum, of which 2–4 % occurred during storage. Losses stood at approximately 15 % for rice, with wide disparities from country to country, notably during storage.

Roots and Tubers

Roots and tubers (cassava, yams, potatoes, sweet potatoes, etc.) are also a dietary staple in many developing countries, notably sub-Saharan Africa. They are bulky and have a high water content (up to 80 %). Post-harvest losses can be accounted for by the following: damage during mechanical harvesting processes or subsequent handling, physiological deterioration, damage caused by insects and rodents during storage, and losses incurred during industrial processing. For cassava, total losses can reach 45 % with traditional agricultural systems [11]. For yams, the staple food of African countries of the Gulf of Benin (up to 30 % of calorie intake) and prized as fresh produce in urban settings, cumulated post-harvest losses can reach 50 % under the combined influence of metabolic activity, germination, attack by insects and rats, and crushing during transport [12, 13]. These losses of fresh produce and the seasonality of yields can lead to volatile market prices. An industrial processing, preservation and distribution system was therefore designed to remedy this instability: cassava chips, which are particularly intended for urban consumers [14].

Table 4.3 Examples of PHL profiles according to climate, crop and scale of farming (percentage of total) [10]

Climate cluster[a]	A	C	B	B	A
Crop	Maize	Maize	Sorghum	Millet	Rice
Scale of farming	Small-scale	Large-scale	Small-scale	Small-scale	Small-scale
Harvest and field drying	6.4	2	4.9	3.5	4.3
Drying	4	3.5	–	–	–
Shelling/threshing	1.2	2.3	4	2.5	2.6
Winnowing	–	–	–	–	2.5
Transport to store	2.3	1.9	2.1	2.5	1.3
Farm storage	5.3	2.1	2.2	1.1	1.2
Transport to market	1	1	1	1	1
Market storage	4	4	4	4	4
Cumulated losses	17.9	11.3	12.6	9.3	11.4

[a] Climate codes: A: Tropical savanna; B: Arid/desert; C: Warm temperate

Fruits and Vegetables

There is a paucity of data for post-harvest losses of fruits and vegetables. More than for cereals, roots and tubers, data for fruits and vegetables depend on the product. Certain tropical fruits, for example papaya and avocado, are particularly fragile (losses of 30–60 % for papaya). A lack of consistency in the methods used to measure losses is an impediment to accurate comparisons. Some experts consider that losses are greater in developing countries [15, 16], still others claim that there is no significant difference [17].

Post-harvest Losses and When They Occur

The majority of developing countries are situated in the world's tropical zones. Some must cope with both high temperatures and humidity levels, which exacerbate the causes of post-harvest losses, e.g. the proliferation of micro-organisms and insects—Grolleaud [4] cites a 50-fold monthly proliferation rate for the bean weevil (beetle that thrives primarily in the seed) and 70-fold for *tribolium* (flour beetle that also attacks other cereal crops)—and physiological and biochemical ripening, deterioration and germination processes of crops. These factors come into play particularly during storage. Losses at this level have been the subject of most studies and the determining factors have been clearly identified. A distinction is made between inherent factors which affect the natural evolution of living produce and therefore metabolic reactions, external physical factors (rough handling, temperature, humidity, etc.), and biological factors (moulds, insects, rodents, etc.).

The two main factors that alter food stocks, i.e. humidity and temperature, are used to distinguish humid tropical areas from arid tropical areas. In the case of humid tropical areas, preservation is a challenge due to the difficulty of drying, and therefore the risk of mould (mycotoxins). In dry tropical areas, where raw commodities are usually stored after the rainy season, predators are the main impediment to preservation (birds, insects and rodents). In arid areas, drying also poses problems during the rainy season, but mainly for processed food products. Before harvest, frequent rains keep the grains moist, making them vulnerable to the development of moulds during maturation in the field. Weather also plays a vital role: drying of standing crops in the field, a common practice among small-scale farmers who often lack appropriate drying structures, is a particularly vulnerable time—lasting 2–3 months—during which grain is exposed not only to humidity but also attack from insects, rodents and birds.

When it comes to cereal grains, moisture levels at harvest are a crucial factor at several different stages: too much moisture and yields must be dried to avoid the risk of heavy losses during threshing and storage; too dry, in the case of rice, for example, and threshing, hulling and processing can lead to breakage and damage that make the crop more vulnerable to attack from certain insects [4]. Moreover, excessive humidity levels coupled with high temperatures promote the development of moulds that are responsible for mycotoxin contamination. Mycotoxins are secondary toxic metabolites produced by certain moulds on a large number of food crops (cereals, ground nuts, walnuts, coffee, cacao, grapes, spices, etc.), in the field and after harvest, but especially during storage. According to the FAO figures, 25 % of the world's crops are contaminated by mycotoxins [18]. Perishable produce (roots and tubers, fruits and vegetables, meat and fish) is particularly prone to losses, both in terms of quantity and quality, because it is more likely to be damaged during mechanical handling, and therefore to deterioration caused by micro-organisms, quickly making it unfit for consumption, notably in high temperatures of tropical zones. All products taken together, the determining factors of post-harvest losses in developing countries are numerous and can come into play at any stage of the harvest and post-harvest chain, as shown in Table 4.4.

Next Steps

From the 1970s and for many years, the main strategies carried out by the FAO and other agencies, development services and research centres focused on battling pests by chemical means. Many projects involved modernizing on-farm, centralised storage facilities. A significant effort was made to train personnel involved in making information accessible to the general public, notably via manuals. Forty years on, what assessment of these efforts can be made? Four main lessons can be drawn from the 2010 workshop organised by the FAO and the World Bank, "Reducing

Table 4.4 Primary and secondary factors causing post-harvest losses and when they occur in harvest and post-harvest systems. (Source: based on CIRAD studies)

Primary factors leading to losses		
Nature of losses	Position in the post-harvest system	Examples
Quantitative		
Accidental	Harvest, transport, handling	Dropped or torn bags, spillage
Due to handling with tools	Harvest, threshing, transport, storage, processing	Breakage of grains
Damage caused by birds	Pre-harvest drying In-field drying of standing crops	
Damage caused by rodents	Drying, transport, storage	Rats, mice
Damage caused by insects	Drying, transport, storage	Larger corn borer
Qualitative		
Physical conditions	Harvest, drying, storage	Heat, cold, humidity
Traces of birds and rodents	Drying, storage	Excretions, feathers, hair
Damage caused by micro-organisms	Drying, storage	Aflatoxin contamination
Traces of insects	Drying, storage	Excretions, larvae, nets
Respiration and transpiration	Storage	Perishable products
Handling	Throughout entire chain	Bruising leading to rot
Secondary factors leading to losses		
Inadequate technical expertise (harvest, packaging, all handling, poor application of recommended technological innovations)		
Lack of suitable tools, storage equipment on and off the farm, transport containers, packaging		
Inadequate means of transport		
Insufficient and/or poor road infrastructure		
Insufficient cold storage		
Lack of drying equipment, poor drying season		
Regulatory quality standards (grading system)		
Unfavourable market conditions (lack of transparency, low prices, etc.)		

Post-Harvest Losses in Grain Supply Chains in Africa", at least for cereal crops in Africa [19].

The first lesson is that the dominant, long-held view according to which traditional agricultural harvest and post-harvest practices are systematically at the root of losses, is a fallacy. The following excerpt from the conclusion of a report on an FAO workshop held in Accra in 1994 to assess African experience in post-harvest techniques serves to illustrate this point [20]: "…it seems that traditional technologies carried out by farmers over the years have proven useful. They are still predominantly in use today because they are well anchored in the culture of the various ethnic groups. Nevertheless, the socio-economic evolution and the ecological changes have considerably altered the production and marketing conditions for staple foods (grain and tubers). The post-harvest system must therefore be adapted to the new context which is continuously evolving." In other words, traditional knowledge, expertise and techniques can be useful and efficient. They do not, however, suffice in an economic, social and environmental context that is constantly evolving.

Moreover, innovations brought in from elsewhere are not always compatible with traditional practices and systems. For example, new varieties of high-yield maize may be more vulnerable to pests because their ears are less protected by husks. Innovations that are not suited to traditional practices can in fact lead to considerable losses: for example, overly frequent and indiscriminate opening of hermetic containers distributed for storing grain in a modified atmosphere, which leads to reoxygenation.

Lesson number two: it is not enough to implement innovations at specific points of the food production chain in an isolated manner. Rather, a comprehensive overview of the entire food production chain must be adopted, taking all players into account, the relationships between them and their interests, which must be reconciled. This way, innovations are more likely to be adopted. This second lesson is illustrated by Insert 4.1. which outlines the POSTCOSECHA project in Central America.

Lesson number three: small-scale farmers will not invest in boosting production beyond their personal needs unless there is something in it for them. In light of this, any action that would lead to higher revenues for growers, less volatile market prices and easier and prolonged access to markets is in order. The same goes for actions geared toward boosting knowledge and expertise related to the storage, processing and sales and marketing of agricultural produce. Along similar lines but from a broader perspective, in addition to these investments, an efficient and stable institutional environment is vital. Without this, no market can function efficiently. The respective roles of the public and private sectors must be clearly identified.

Last but not least, lesson number four is relative to research and development, which must prioritize finding concrete solutions that reconcile modern technology

Insert 4.1. Lessons Learned from the POSTCOSECHA Project in Central America

The POSTCOSECHA project was implemented in 1980 as part of the strategy for rural development aid implemented in Central America by the Swiss Agency for Development and Cooperation. The project's goal was twofold: reduce post-harvest losses of staple foods (maize, beans, millet and rice), and generate extra revenue for local populations. The project was initially rolled out in Honduras in 1980. Once feedback was collected, it was extended to three neighbouring countries: Guatemala in 1990, Nicaragua in 1992 and El Salvador in 1994.

Justification for the project was found in the fact that in all four countries, 16 % of the 3.3 million t of maize, beans, millet and rice harvested were lost during storage due to pests, mould and damage caused by birds and rodents. Depending on product and place, losses reached 27 % for maize, 23 % for beans and 30 % for millet [21].

The project was based on a four-point strategy: (1) simple, efficient and accessible technology acceptable to local populations; primarily metal silos which, when properly used, can eliminate losses; (2) on-site manufacturing of metal silos and transfer of skills to growers by developing a market between them and artisans; (3) an economic model based on a public/private partnership; (4) management and assistance provided by outside third parties (non-governmental organisations, local associations) with the participation of all stakeholders (banks, municipalities, artisans, agricultural services, agricultural education establishments). The project was a success, despite the modest tonnage involved. In 2005, some 500,000 metal silos were used to protect 45,000 t of produce for an annual value of approximately 10 million US $. According to H. Hermann [22], the benefits of the project were reaped at several levels:

- local growers and their families, whose situation improved thanks to increased food availability and improved quality thereof, improved hygiene in and around the home, lighter workload (especially for women), but also more favourable market conditions: a prolonged storage period where losses are eliminated protects growers from volatile market prices and affords them a longer selling period and the possibility of higher prices for their goods;
- artisans, who benefit from higher incomes thanks to the manufacture of silos (creation of some 200 jobs at this level);
- all local communities, who benefit from these advances in the private sector which energize the economy.

According to the FAO [23], metal silos have been distributed and used in 16 countries throughout three continents over the past 10 years.
(to find out more, visit www.postcosecha.net)

and traditional practices (in order to ensure their acceptability) for processing food products. The focus must be primarily on small and medium enterprises and women's associations, which play a key role in food crop processing systems.

An Urgent Need for Technological Innovation

There is one priority directly linked to lesson number four that emerges clearly: the urgent need for technological innovation. Solutions should be specific and tailored to situations where a need for improvement has been identified. Nevertheless, three needs across the board have been defined. First, the need for drying and processing equipment (drying by solar energy, as access to electricity is rare in many developing countries and regions, and access to an alternative renewable energy source of the utmost importance, and parboiling and pre-cooking of processed products, processes for which solar energy can also be of tremendous use). Second, storage equipment such as metal silos (see Insert 4.1.), both at individual level (farms) or at a broader level (group of farms), hermetic storage and transport containers, or better packaging techniques, notably for perishable foods. Lastly, efficient, safe and hygienic methods to protect produce, including traditional knowledge and expertise and integrated protection and production methods, such as the biological control of the larger corn borer (*Prostephanus truncatus*), responsible for considerable damage to maize crops (see Annex at the end of the Chapter). All of these methods should be made available at reasonably moderate prices. Analysis has revealed the need for technological solutions that are simple, affordable, easy-to-uphold and adapted to specific problems—this last criterion being at odds with the first three. They all highlight the importance of targeting women and integrating proposed solutions in the broader framework of each sector according to the product, taking all stakeholders into account. The success of the collective approach has been highlighted, as has that of projects implemented by local populations.

A Favourable Context is Key to Success

As useful as they are, technological solutions alone are not enough to reduce post-harvest losses in a significant and lasting way. They can, however, have a greater impact if they are implemented in an appropriate environment that is conducive to an efficient sector and market. It all comes down to several key priorities. The FAO estimates that investments of up to 940 billion US $ are needed by 2050 to fight hunger in sub-Saharan Africa alone. Approximately half of that must be geared toward a four-fold strategy to reduce post-harvest losses: cold and dry storage, road infrastructure in rural areas (particularly important to access collective storage facilities and markets), open-air markets and wholesale trade, and lastly the initial processing stage of raw commodities.

Substantial investments are needed, but the question of financing remains. As things currently stand, funding is notoriously lacking. Development banks and donors, both international and national, have yet to make reducing post-harvest losses a priority. The banking sector's role is all the more vital given that heavy investments are needed, and expected returns on investment can take up to 10–15 years. Clearly, microcredit will play a limited role. In such a context, some interesting initiatives have been taken, such as financing agricultural produce by warehouse receipt financing on inventory credit (see Insert 4.2).

Farmers' organisations and associations also have a key role to play, as many examples with very different circumstances have shown. Their influence in promoting good practices has been illustrated in an association of banana farmers in India who sell their produce directly on the market [24, 25]. As part of a common framework of requirements, this association limits post-harvest losses by recommending the following: careful loading of fruit, limited handling, improved transport, etc. The result is reduced post-harvest losses (approximately 18 %) compared with alternative systems that are not a part of farmer associations and that deal in wholesale (approximately 29 %).

Insert 4.2. A Role for the Private Sector: Warehouse Receipt Financing on Inventory Credit

Few countries and sectors can rely on the complete commitment of their governments, and agriculture is no exception. Government bodies, which formerly saw to purchasing and storing harvests between growing seasons and lean periods, often eschew this duty, leaving the sector to its own devices to find new ways of doing business if the agricultural sector is not to be subjected to market forces alone. The private sector does not have easy access to the necessary credit to finance agricultural trade, as this comes with no guarantee in the majority of developing countries. The FAO and NRI have suggested warehouse receipt financing on inventory credit—a method of gaining access to money with a long history—to remedy the problem. This involves three players: growers, warehouses and banks. The grower stores his harvest, in whole or in part, in a warehouse in exchange for a certificate of receipt. This receipt acts as a guarantee, allowing the grower to then get a loan from the bank. Coulter and Shepherd [25] have studied several examples of this system currently in use in various countries, analysing the advantages and disadvantages and conditions for success. In many cases, the system has been found to work very well.

Lastly, if growers are to invest in post-harvest preservation/conservation equipment, adopt good practices and gain access to the market, they must be better trained. However, growers are not the only ones in need of proper training; programmes must be extended to include ministries and agricultural services as well. The World Food Programme's 2009 Purchase for Progress initiative, or P4P, bears witness to the importance of this fourth course of action. As its name suggests, the initiative is designed to help farmers market and sell their cereal crops via different types of contracts: direct contracts for smallholder farmers, contracts with local traders, and 3-year contracts. These contracts, which are advantageous for farmers in terms of price, impose higher quality standards for grain crops, in particular low mycotoxin content. To achieve this, the WFP has teamed up with the FAO and NGOs to provide training in post-harvest conservation methods and supply farmers with processing equipment. Incidentally, this is yet another example of how an integrated, multi-level approach gets better results, in this case focusing on training, equipment and sales contracts.

Throughout history and in all civilisations, organic waste and produce unfit for human consumption have been used to feed animals, be it pets or livestock. As seen in the last Chapter, it is during the distribution and consumption stages that this comes into play, but it may also be a means to put post-harvest losses to good use. Animal husbandry is part of many traditional farming systems in developing countries, via aquaculture or small- or large-scale husbandry. Farmers who practice animal rearing exceed the 1 billion mark. The guiding principle of these systems is that there is no such thing as definitive waste. Waste from a given agricultural yield is put to use in one or more other agricultural productions. The so-called VAC system, used in South-East Asia, notably Vietnam, is undoubtedly the most telling example. VAC stands for "*Vuon*" (vegetable garden, but in a broader sense, all the plants cultivated in a garden, field or forest), "*Ao*" (pond, or more broadly, all aquaculture activities, be it fresh or salt water) and "*Chuong*" (animal pens, by extension all activities related to animal raising, including bird- and beekeeping). The most common VAC system involves growing vegetables, fish farming and raising pigs, but sometimes pigs are replaced with ducks, rabbit, goats, poultry, etc. The VAC system requires water, which therefore must be available in sufficient quantities. Compared to other traditional systems, where the three types of production are carried out separately, it has been shown that the VAC system is, overall, more productive per unit of surface area, more economical when it comes to using resources, and more environmentally-friendly [26]. Another benefit of the VAC and related systems is that they provide households with a more varied diet, and therefore contribute to improving the health of malnourished populations [27]. Nevertheless, there are drawbacks: they require large amounts of water and therefore cannot be implemented in areas where water is scarce. They also come with a high degree of risk in terms of hygiene and safety (microbiological and epizootic), linked to intense animal farming.

Beyond this example, the point of focusing on the advantages and the disadvantages of the VAC system is to illustrate that the issue of losses—post-harvest losses in developing countries, as discussed in this Chapter—must be put in perspective and analysed within a larger context. We must look at the big picture in terms of: all phases of the vertical food supply chain, from farm to fork; integrating the spatial dimension and associations of different activities in the same area—the merit of the VAC system lies in the fact that it aims to reduce post-harvest losses by combining crops and animal rearing in a circular agricultural system; what is true for strictly agricultural activities also holds for the combination of several activities, agricultural or not, production-oriented or not (home, recreation, etc.) within the same area; not limiting the scope to the present but adopting a dynamic, long-term vision; and lastly, applying analyses at the local level to global food, energy and environmental issues. To reduce post-harvest losses in developing countries, action must be taken at several levels at the same time, as discussed in this Chapter. Although it will take considerable effort (in terms of measuring losses and identifying their causes in different situations throughout the world), experts know relatively well which areas need attention. As is the case all too often, it all comes down to money. In other words, beyond the scientific, technical and training aspects, if post-harvest losses in developing countries remain a major challenge today, it is because investments are severely lacking.

Annex

Estimation of weight loss of stocks (in %) for different grain crops in different climates, adjusted to 9-month storage periods and the average consumer household [28].

Species/Climate	Scale of farming	Variety	Estimated loss
Ears of maize with *Prostephanus truncatus* infestation			
Aw	Small-scale	Local	9.7
Bsh	Large-scale	Local	2.7
	Small-scale	Local	13.3
Cwa/Aw	Large-scale	Local	2.1
	Small-scale	Local	10.0
Ears of maize without *Prostephanus truncatus* infestation			
Aw	Small-scale	Local	5.3
Bsh	Small-scale	Local	4.3
Cwa	Small-scale	Local	3.5
	Small-scale	Hybrid	9.5
Maize kernels with *Prostephanus truncatus* infestation			
Aw	Small-scale	Local	5.4
Bsh	Small-scale	Local	3.3
Aw	Small-scale	Local	5.4
Cwa	Small-scale	Local	4.2
Sorghum seeds			
Bsh/Bwh	Small-scale	Local	2.5
Cwa	Small-scale	Local	3.9
Sorghum panicles			
Aw	Small-scale	Local	2.8
	Small-scale	Improved	11.0
Millet			
Bsh	Small-scale	Local	1.1
Cwa	Small-scale	Local	1.3
Wheat			
Bsh	Small-scale	Local	3.1
Cwa	Small-scale	Local	5.8
Barley			
Cwa/Cwb	Small-scale	Local	0.8
Rice			
Aw	Small-scale		1.2
Bsh	Small-scale		0.1
Cwa	Small-scale		0.4
Teff			
Cwa	Small-scale	Local	0.3

Definition of climate zones: Aw: tropical savanna; Bsh: semi-arid/arid; Bwh: hot desert; Cwa: temperate with dry winters and hot summers; Cwb: temperate with dry winters and mild summers

References

1. Vignau-Loustau L, Huyghe C (2008) Stratégies fourragères: Pâturage—ensilage—foin. France Agricole, 336 p
2. Dulphy JP, Martin-Rosset W (2001) Valeurs comparatives de différents types de fourrages conservés. In: Garel JP, Josien E (eds) Étude de la faisabilité de la mise en œuvre de l'arrêt de l'utilisation des fourrages fermentés. Pôle fromager AOC Massif Central, Aurillac, pp. 12–30
3. Demarquilly C, Dulphy JP, Andrieu JP (1998) Valeur nutritive et alimentaire des fourrages selon les techniques de conservation: foin, ensilage, enrubannage. Fourrages 155:349–369
4. Grolleaud M (1997) Pertes après récolte: un concept mal défini ou mal utilisé. Étude synthétique et didactique sur le phénomène des pertes se produisant tout au long du système après récolte. FAO, Rome
5. http://www.fao.org/GIEWS/english/tools.htm#m
6. Tyler PS (1982) Misconception of food losses. Food Nutr Bull 4(2):21–24
7. Harris KL (1978) Postharvest grain loss assessment methods: a manual of methods for the evaluation of postharvest losses. American Association of Cereal Chemists, 193 p
8. Schulten GGM (1982) Post-harvest losses in tropical Africa and their prevention. Food Nutr Bull 4(2):1–24
9. www.phlosses.net
10. http://www.phlosses.net/downloads/Users_Guide_version_1_6.pdf. Accessed 14 April 2011
11. Jeon YW, Halos LS (1991) Addressing Research and Development for Cassava Postharvest System in West Africa. International Winter Meeting, The American Society of Agricultural Engineers, Chicago
12. Berthaud J, Bricas N, Marchand J-L (eds) (1998) L'igname, plante séculaire et culture d'avenir. Actes du séminaire international Cirad-Inra-Orstom-Coraf, 3–6 juin 1997, Montpellier, France
13. Osunde ZD (2008) Minimizing postharvest losses in yam (Dioscorea spp.): treatments and techniques (Chapter 12). In: Robertson GL, Lupien JR (eds) Using food science and technology to improve nutrition and promote national development. International Union of Food Science & Technology
14. Vernier P, Bricas N, Ategbo EA, Hounhouigan J, N'Kpenu K, Orkwor G (1999) Le développement des transformations post-récolte pour l'approvisionnement des marchés urbains: l'exemple de la filière cossette d'igname en Afrique de l'Ouest. In: Borgemeister C, Bell A, Mück O, Zweigert M (eds) De la lutte biologique à une approche systèmes de la post-récolte. Compte rendu d'une réunion IITA-GTZ, du 13 au 15 octobre 1997, Cotonou; Eschborn, GTZ, pp 169–183
15. Subramanyam KV (1986) Post harvest losses in horticultural crops: an appraisal in India. Agric Situat India 41:339–343
16. Coursey DG, Booth RH (1972) The post-harvest phytopathology of perishable tropical produce. Rev Plant Pathol 51(12):751–765
17. Kader AA (2005) Increasing food availability by reducing postharvest losses of fresh produce. In: Mencarelli F, Tonutti P (eds) Proceedings of the 5th International Postharvest Symposium. Acta Hort. 682, ISHS 2005
18. http://www.fao.org/ag/agn/agns/chemicals_mycotoxins_en.asp. Accessed 15 April 2011
19. FAO and World Bank (2010) Reducing post-harvest losses in grain supply chains in Africa. 18–19 March 2010, FAO, Rome. http://www.fao.org/ag/ags/publications/docs/misc/FAO_WB_ph_web.pdf
20. FAO (1994) Synthèse de l'expérience africaine en amélioration des techniques après récolte. Journées techniques Accra, Ghana. Service du Génie Rural AGSE. FAO, Rome, Italie. See also http://www.fao.org/docrep/w1544e/w1544E07.htm
21. Schneider K (1997) Grain silos for everybody—"PostCosecha" in Central America: the story of a successful project for post-harvest technology. Report: available on: http://www.postcosecha.net/en/Home/Documentation/Publications/Publications_in_English

22. Herrmann H (1991) Seguridad alimentaria: Comparación de impactos socio-económicos en la tenencia del silo metálico versus sistema tradicional de almacenamiento. Research report. Summary in English available on: http://www.postcosecha.net/en/Home/Benefits_Impact
23. FAO (2008) Les silos métalliques familiaux: alliés essentiels de la FAO dans la lutte contre la faim. http://www.fao.org/ag/ags/publications/docs/misc/silos_F_light.pdf
24. Murthy DS, Gajanana TM, Sudha M, Dakshinamoorthy V (2009) Marketing and post-harvest losses in fruits: its implications on availability and economy. Indian J Agric Econ 64(2):259–275
25. Coulter J, Shepherd AW (2001) Le crédit sur nantissement des stocks—une stratégie de développement des marchés agricoles. Bulletin des services agricoles de la FAO 120
26. Kumar K, Ayyappan S (1998) Current practises in integrated aquaculture. Central Institute of Freshwater Aquaculture, Working Paper Number 5
27. Hop LT (2003) Programs to improve production and consumption of animal source foods and malnutrition in Vietnam. Conference "Animal Source Foods and Nutrition in Developing Countries", Washington, June 24–26, 2002
28. Hodges R (2004) Crop post-harvest: science and technology. In: Hodges R, Farrell G (eds) Durable case studies in the handling and storage of durable commodities, vol 2. Blackwell Publishing Company, 288 p

Chapter 5
Producing Other Goods

In addition to being a source of food, agriculture has long been the primary—if not exclusive—source of energy, medicine, dyes, textiles and other materials useful to mankind. This role has weakened over the past century in industrialised countries with the development of chemical synthesis. As a result, agricultural production has focused on supplying food goods, even if there are large industrial niches that do not supply food, but rather raw materials such as cotton, sisal, flax, wool, silk, leather, rubber, gums, paper, wood and woodchips, and starches.

As petroleum, a necessary energy input and petrochemical raw material, becomes increasingly scarce in a world with expanding demographics and economies, and where global warming is an ever-present concern, action must be taken quickly. Demand must be mitigated by changing lifestyles and consumption habits that use excessive amounts of energy and non-renewable products; supply must be supplemented by finding alternatives to fossil fuels. New efforts are being directed at finding non-food uses for agricultural products, and more specifically, for biomass.

What Is Biomass?

An Abundant Resource

Biomass refers to all the biological matter of living organisms and their production. Plant, terrestrial and aquatic biomass result from photosynthesis. Biomass is carbon-based matter produced by plants that extract carbon from carbon dioxide in the atmosphere. These plants draw their energy from the sun and resources in the ground or surrounding environment, such as water, nitrogen, potassium and phosphorus. Although biomass is a renewable raw material, a certain number of non-renewable resources are necessary for its production, such as petrochemical nitrogen-based fertilisers or those developed using potassium and phosphate extracted during mining. According to Wirsenius [1], total annual use of terrestrial biomass for food and non-food end uses represented 13.4 billion t of dry matter produced on 7.7 bil-

M. Guillou, G. Matheron, *The World's Challenge*,
DOI 10.1007/978-94-017-8569-3_5,

lion ha (Bha) of land. Of these 7.7 Bha, 5.2 Bha (1.6 Bha for crops and 3.6 Bha for pasture) were meant for human consumption, whether directly or indirectly via animal production and feed. Some 2.6 Bha of forests, of which 2.4 Bha were natural or semi-natural, were used to produce wood, including for energy. Only several million hectares were used to produce bioenergy and bioproducts. Despite uncertainty and imprecision, the figures clearly show how food-related agricultural production far outweighs non-food related production, at least until 2005.

Competition between food and non-food uses of agricultural products did not make headlines until recently. New uses were presented as opportunities for growth for those in the agricultural industry. However, the issue came to light because the development of a non-food use—biofuels—coincided with sharp price increases of agricultural products at the end of 2007 and the first quarter of 2008. These price hikes led to riots in many cities across the world and exacerbated world hunger. Except for maize grown in the United States, where 40 % of the crop goes to producing ethanol, no incontestable link was established between increased energy crops and higher agricultural food prices. Nevertheless, events served to raise awareness of the possible conflicts in the end uses of agricultural products, considering the world is battling hunger, with nearly 1 billion people suffering from malnutrition (see Chap. 7).

New Outlooks for Biomass Uses

Besides its traditional uses—namely food-related—biomass is a renewable raw material with myriad possible applications, both energy (bioenergy) and industrial (bioproducts and biochemicals). Bioenergy can be broken down into two basic categories: those used for transport (biodiesel and bioethanol) and those used to generate heat and electricity (solid and liquid biofuels and biogas). Bioproducts include biomaterials, which include biodegradable polymers, or biopolymers, used for bioplastics; fibres and composite materials (agrimaterials); and paper and cardboard. Biochemicals include surfactants, solvents, lubricants and cosmetics. Increasingly-scarce petrochemical fossil fuels (petroleum, natural gas, coal) combined with global warming and the need to reduce greenhouse gas (GHG) emissions, are two long-standing factors that lend greater importance to biomass today.

Given the situation, it is understandable that public policies focus on finding solutions that not only support economic activity and promote renewable energy sources and raw materials, but also push for a gradual, widespread decarbonisation in industry. Biomass is viewed by the world as the resource of tomorrow, one with abundant potential uses. The most compelling of the many possibilities is using biofuels derived from crops. However, this poses threats to food production due to potential conflicts regarding land use, because of competition, either direct (food crops) or indirect (converting grasslands into farmland), with food end uses.

High Expectations for Energy

Given the general situation of petroleum use—dwindling resources, rising prices, concerns over GHG emissions—many countries have adopted public policies to promote the switch to biomass energy. According to the "Climate and Energy Package" approved by the Council of the European Union and the European Parliament in December 2008, all Member States must reduce GHG emissions by 20 % compared to 1990 levels, achieve energy savings of 20 % and increase the proportion of renewable energies to 20 % by 2020. This package extends and broadens European progress made under the Kyoto Protocol.

The aim to replace petroleum with biofuels presents an enormous challenge in terms of production. This is no black and white matter, with bioenergies being "virtuous" because they are produced using CO_2 in the atmosphere. The stakes are high: geopolitical and strategic risks related to energy issues, such as the decentralisation of production; environmental risks linked to the potential impacts along the entire chain of production; and economic and social risks concerning how to organise new sectors and establishing GHG emissions trading schemes. Moreover, from a strictly quantitative standpoint, given the massive energy needs, such large-scale production quickly runs up against the finite nature of available resources—particularly land.

From Firewood to Tomorrow's Bioenergy

Fossil fuels represent nearly 80 % of the world's total primary energy consumption. According to some estimates, oil and natural gas resources could be exhausted by the middle of this century, with global energy demand reaching approximately 20 Gt of oil equivalent (Gtoe; 1 Gt equals 1 billion t) per year, as opposed to 12 Gtoe today. Transitioning towards other energy sources is therefore essential. The proportion of plant biomass in the global primary energy mix has declined in recent years, from 15 % in 1970 to just 10 % in 2005 (4 % in France). However, biomass uses for energy have witnessed sharp growth, in both the forestry and biofuels sectors. In northern countries (Scandinavia, Canada), which have large amounts of forest cover and a well-established forestry industry, collective and industrial uses of ligneous fuels, woodchips, charcoal and forestry by-products have developed. In France, it is estimated that out of 125 million m^3 of wood produced annually in forests, only 50 million are used, which demonstrates that the resource is available. Wood energy represented 46 % of the 19 Mt of oil equivalent (Mtoe) of renewable energies in 2008, with forecasts projected at 41 % of 36.6 Mtoe of renewable energies in 2020. The decisive technological advance was improvement in energy efficiency. While traditional wood burning stoves converted only 35 % of the potential energy in wood, efficiency increases to nearly 80 % with modern wood pellet stoves for home use, rising to nearly 90 % when used with combined heat and electricity systems. This breakthrough is due to development in industrial applications for combined electricity and heat production on both collective and individual scales, with small- and large-sized co-generation systems. Such biomass-run production plants could replace those relying on coal or gas to handle peaks in electricity use.

In Africa, 88 % of wood extraction is used for firewood. In the least-developed African countries, the proportion of populations with access to electricity is—and will remain—marginal for many years to come. In these countries, which are major consumers of wood energy, either in its natural state or as charcoal, consumption of ligneous fuels is on the rise. Consequently, wood is becoming increasingly scarce, especially in very dry regions, and is a large factor in deforestation. Producing and managing forest resources, whether through the use of plantations or natural forests, must go hand in hand with more energy-efficient technologies, as wood charcoal has extremely low energy efficiency; sawdust and sawmill by-products must also be utilised. In Asia, a long-standing tradition of biogas use in homes has led to improved biomass energy efficiency, which couples biomass energy with the use of livestock manure. But in many areas of Asia, as well as in Latin America, deforestation for wood or other reasons—especially to make way for crops—is problematic.

Fuel: A Particularly Demanding Energy Source

For many centuries, animals were the main source of available power. Until the mid-nineteenth century in Europe, animals were raised first and foremost as a source of energy, rather than for milk or meat. Several factors came together to cause a shift in this paradigm: the invention of the steam engine in the eighteenth century, followed by the internal combustion engine in the nineteenth century, and finally the discovery of coal, gas and oil resources. The widespread use of internal combustion engines and the extraordinary development of transport was a fundamental factor in shaping economic development. The availability of cheap fossil fuels led to significant declines in energy prices, which was quite favourable to industrialisation and transport development [2].

Oil is difficult to replace as a fuel for transport—its main application—and the quantities at stake are staggering. Because liquid fuels are a compact, easy-to-use and stable energy source, they will most likely remain the primary source used for transport in the future. This puts biofuels on the fast track to development, provided they are adapted to the specifications of today's combustion engines. Over the past 30 years, Brazil is the only country to have adopted a large-scale policy of substituting biofuels for oil, first using ethanol hydrate, followed by incorporating anhydrous ethanol (between 25 and 35 %) in fuel. This type of ethanol, produced from sugarcane, has helped make Brazil the world leader in sugarcane production: more than 30 % of the land in the world devoted to sugarcane production is in Brazil. The United States has followed suit, incorporating maize ethanol into fuel, and many European countries have ambitious plans to do the same with diesel or regular fuel, depending on the car fleets in question.

Despite this spike in recent years, world production of biofuels has remained low compared to oil production. In 2008, it was 43 Mtoe, or slightly over 2 % of all fuels used in road transport, and most was in the form of bioethanol in the United States and Brazil or biodiesel in the European Union. This required some 320 million t of sugarcane (17 % of global production), 100 million t of cereals (5 %) and 11 million t of vegetable oils (9 %). Furthermore, 28 million ha of land were used, equalling 3 % of global oleaginous, cereal and sugarcane crop lands [3]. Although modest, this production has seen rapid development since the early 2000s and has sped up in recent years. In 2008–2009, increased world demand for cereals for bioethanol production (28 million t) exceeded demand for food uses (13 million t) [3].

These figures clearly show the principal drawback of biofuels: their voracious appetite for land with regards to current technology and energy production efficiency and direct competition of cereal, oleaginous and sugarcane plants for food uses. Many are hopeful that more efficient biofuels, or those that do not compete directly with food uses, can be developed. Already, the first generation of biofuels has been replaced by a second, which is sure to be replaced by even more evolved generations in the future.

First- and Second-generation Biofuels

First-generation biofuels use plant storage organs, either directly, as with oleaginous plant oils, or after fermentation, such as cereal starches after hydrolysis and cane or beet sugars; the technologies used to produce them are mature. Experience gained from these technologies can be exploited to move towards a second generation of biofuels. First-generation biofuels use only part of the plant (except for sugarcane in Brazil where the parts not used in ethanol production are used to produce the energy necessary for processing cane sugar); this results in many by-products, most of which are used for animal feed. Such by-products must be taken into account in any analysis aiming to provide an economic or environmental assessment of current biofuels. They also contribute to the competitiveness of the animal sectors that use them.

Second-generation biofuels use plants in their entirety. They are made from lignocellulose, which is made up of varying amounts of lignin, hemicellulose, and cellulose. Lignocellulose is the main component in secondary plant cell walls. As such, it is the most abundant biomass component in the world [4]. It has a fibrous structure, with its molecules organised into polymers. Lignin is the component that prevents the depolymerisation of plant matter. Steam explosion, acidic and alkaline pretreatments enable the lignin to be separated out and facilitate the saccharification of cellulose and hemicellulose. There are two types of biomass conversion processes that can be used to obtain second-generation biofuels: thermochemical and biochemical conversion. Thermochemical conversion involves heating biomass to break down the molecules (e.g., gasification to obtain syngas), followed by catalytic conversion to obtain diesel- or kerosene-type hydrocarbons. This type of conversion faces certain challenges, including ageing catalysers and the difficulty of obtaining the right gas mixture to be used in gasification. Biochemical conversion produces alcohols such as ethanol or butanol. Once the raw material has been broken down, the complex sugars making up the lignocellulose are broken down into simple sugars via hydrolysis. The resulting glucose and other simple sugars are then fermented to become ethanol; finally, they are separated through distillation to remove sediment. The technological challenge in this type of conversion lies in conditioning, or making cellulose and hemicellulose suitable for enzymes to be able to break them down into simple sugars, which later undergo fermentation by microorganisms. The main difficulties reside in the accessibility of the monomers, i.e., hexose and pentose, and the cost of the enzymes. While thermochemical conversion requires large installations and hefty investment to take advantage of economies of scale, biochemical conversion can make use of existing installations already in place for first-generation biofuels. Both, however, require steep investment to move beyond the laboratory and field demonstrations to be applied at an industrial scale.

There are three main biomass sources which can be exploited for second-generation biofuels: residues and waste from various sources, including agriculture (cereal straw, maize stalks, rapeseed stems), forestry (branches, stem wood and cull trees), industry, urban areas and individual homes; forest resources, such as wood from forest stands; and dedicated crops, whether annual crops (use of the whole plant—wheat, maize, etc.), perennial grazing plants (fescue, orchard and rye grasses), herbaceous plants (Miscanthus, also known as Elephant Grass, and switchgrass) or shrubs and trees (poplars, willows, eucalyptus and locusts, which are kept pruned short and harvested every 3–10 years). Considerable research has been carried out with a view to identifying and quantifying various usable biomass types, the cost of their use, and the adaptation of technologies to structural specificities of each source.

Sustainability Must Remain a Priority

In addition to considerations of land use competition (Chap. 7) and the fact that cropping systems must be both productive and ecological (Chap. 8), ensuring the sustainability of the biofuel chain of production is essential. There are many varied technical solutions for producing biofuels, and each one of them should focus on being sustainable. The Millennium Ecosystem Assessment [5] showed that man has accelerated ecosystem changes over the past 50 years in his quest to meet growing needs, causing significant and often irreversible losses in biodiversity, especially in agroecosystems to meet rising food needs. Certain ecosystems, such as natural and semi-natural forests, fragile areas, wetlands, marshes and grasslands, should not be converted to cultivated land so that they can maintain their biodiversity and utility as carbon sinks. This leads to a first criterion for evaluating biofuels: tendencies in local and global land uses, particularly the development of energy crops, which may displace certain food crops or threaten zones that should be protected. Various studies have shown that converting land to energy crop use can result in massive GHG emissions and radically alter the environmental assessment of biofuels [6]. For instance, under certain climatic and research conditions, converting 1 ha of meadow into cultivated land for the production of a first-generation biofuel can release up to

300 t of CO_2 per hectare; converting 1 ha of forest land can release between 600 and 1000 t of CO_2 [7]. Furthermore, the environmental impact is not limited to GHG emissions. Potential negative effects of converting land to be used for cultivation must also be considered, such as losses in biodiversity or fertiliser and pesticide use that exceeds what the natural environment is able to handle. This is especially pertinent when the land is used intensively, i.e. when large quantities of water, fertiliser or pesticides per land unit are required.

The second criterion concerns reducing GHG emissions relative to fossil fuels. In the European Union, only those biofuels which enable GHG emissions to be reduced by at least 35 % compared to petrol or diesel are considered sustainable. That percentage will increase to 50 % in 2017 and 60 % in 2018 for biofuels produced in new fuel conversion plants. These GHG emissions are counted as the sum of emissions from crop production, biofuel conversion and transport; they do not take into account GHG emissions linked to potential conversion of land from one use to another, which is a major factor to keep in mind. Under such restrictive conditions and compared to carbon fossil fuels, all biofuels currently being used result in reduced CO_2 emissions, ranging from 10 to 15 % for maize in the United States, 40–59 % for rapeseed and 38–59 % for sugar beets in the European Union, 49–80 % for oil palm, and 69–88 % for sugarcane in Brazil. While the environmental impact of nitrous oxide (N_2O) from these energy crops is often poorly understood, what is clear is that the greenhouse effect of one unit of N_2O is 300 times greater than one unit of CO_2. As far as second-generation biofuels are concerned, established forecasts using laboratory research or field demonstrations show reduced CO_2 emissions of 75 % compared to petrol and 90 % compared to diesel.

The third criterion concerns biofuel energy efficiency. Objectives aiming to reduce dependence on fossil fuels and replacing them with renewable energies are essential driving forces behind biofuel development. Net contribution of biofuels to the energy supply depends both on energy content itself and on the energy that is used throughout production and distribution processes, both direct (tractors, lorries, factories) and indirect (fertilisers, pesticides, electricity). The energy assessment can vary greatly depending on the crops, production systems and sectors. On this point, biofuels from tropical biomass such as sugarcane and oil palm present a clear advantage in comparison to first-generation biofuels from other biomass sources. These energy assessments are often a point of contention in terms of evaluation methods, especially concerning how analysis criteria are established and how by-products are considered. For example, energy savings gained using a rapeseed-derived biofuel compared to diesel can vary from 26 % to over 70 % [8]. Certain authors, who happen to be highly controversial, contend that the energy balance for certain first-generation biofuels may even be negative, e.g. soybean and sunflower biodiesel or maize bioethanol. As is the case for environmental impact assessments, the way energy balances are calculated, and in particular, Life Cycle Assessment methods, need improvement in order to better assess and manage new energy sectors. Concerning this point, it is also important that the energy balance of oil not be considered as set in stone; due to oil's increasing scarcity, reserves are ever more costly to utilise, both economically and energy-wise.

Finally, the economic impacts of biofuels in many parts of the world must be evaluated, especially with regards to policies that have been adopted, particularly in the United States and the European Union, to promote biofuel development. Generally speaking, such policies seem rather ineffective in terms of public taxpayer money spent compared to fewer total benefits derived for the overall economy. The positive effects for those receiving aid—chiefly farmers—are overshadowed by the negative consequences for others, mainly the final consumers and taxpayers [9, 10]. It should be noted that these economic analyses are limited to private costs and benefits. They do not consider the monetary value of tonnes of carbon averted—a point which goes back to reducing CO_2 emissions where possible—and do take into account potential changes regarding land use. Economic analyses are also highly dependent upon current prices of the various biomass sources used as raw materials for biofuel production.

New Possibilities for Green Chemistry

The second major stake in expanding biomass production is related to chemistry-oriented technology transfers, with the hope of phasing out traditional petrochemistry.

A New Concept: Biorefinery

Over the past several years, the exploitation of non-food agricultural products has gradually moved towards an integrated approach to biomass and to designing new production and processing systems—biorefineries—to manufacture bioenergies and bioproducts. This change is taking place within a radically different context. The twofold global challenge of increasingly scarce petroleum resources and the need to reduce GHG emissions has certainly played a part. But it is also due to the development of "green chemistry", defined in 1991 by the Environment Protection Agency (EPA) in the United States, with 12 principles outlined by American chemists Paul Anastas and John Warner [11]. Biofuel production seemed to be one of many ways to put biomass to use. The concept of biorefinery is characterised by:

- the total exploitation of biomass input;
- mainly using biotechnologies to benefit from greater selectivity and decreased costs associated with these processes compared to chemical processes;
- optimised management of water, liquid and gaseous outflows, and energies;
- optimised management of end products, with by-products being classified as chemical commodities (fuels, intermediate chemical products and chemical materials).

Biomass fractionation and pretreatment is a critical issue in biorefinery processing, whether a matter of expanding the range of possibilities or output. Due to the wide diversity of directions envisioned, the extent of industrial solutions depending on heavy industry and the intense debate in society, an analysis based on multiple criteria, from the "cradle to the grave", needs to be established to ensure processes that are better overall. This type of approach enables all pollution that is transferred between players or areas to be identified, while also identifying points that can be improved on a priority basis according to the different stages of each business sector involved [12, 13]. This systemic approach must also take into account opportunities to integrate several different business sectors on one industrial site or supply basin.

The Role of Synthetic Biology in Renewable Carbon Chemistry

In addition to energy, food and the environment, modern societies require many other services, such as clothing, housing and hygiene. Myriad substances are required to supply such services; fossil fuel derivatives account for a sizeable part. The challenges facing bioenergies that have been previously discussed are also valid for these different types of services. In France, commitments made during the Grenelle Environment Forum are meant to achieve 17 % of biological carbon use in chemistry by 2015 as opposed to just 7 % today. Other countries are even more ambitious: the Netherlands hopes to reach 30 % by 2030. When compared to fossil fuel products, biochemicals and biomaterials can have certain advantages, such as biodegradability or only slight toxicity, which, given current trends in regulations and society's expectations, could encourage their further development. There are two differences concerning bioenergies which must be considered: the volume of matter at stake and the diversity of the molecules involved. Substituting fossil fuels for renewable carbon means no longer working on identical molecules, but on molecules that perform the same functions. Another advantage is that renewable carbon chemistry does not exclude classic organic chemistry. The technological challenge lies therefore in skilfully combining biological and chemical processes based on the four criteria previously outlined for biofuels: environmental impact of required plant production, reduced GHG emissions over the entire life cycle, functional effectiveness of the molecules and economic viability.

Given the situation, all white biotechnology tools should be used, from enzymes to groundbreaking synthetic biology technologies. Such technology is different to traditional metabolic engineering because right from the start of analysis, it takes metabolic processes and necessary enzymes into account. With enzyme remodelling, reactions could be created that are unseen in nature. This considerable technological potential, which could feasibly provide solutions to the various challenges discussed previously, is the point at which biology, physics and control engineering come together. This type of synthetic biology requires modelling tools in order to integrate the different spatial and temporal scales of overlapping reactions. It approaches systems biology, which deals with natural beings. It could be used in second-generation cell factories, and will require sharply defined objectives and practices.

Meeting These New Needs by Choosing the Right Biomass

Plant Sources

Bioenergies and bioproducts are derived from biomass, its greatest strength being that it is renewable, needing only human effort and sunshine. This advantage, however, does not outweigh the fact that it must be processed and put to use. Thus, the development of plant genome analyses will better enable biosynthesis pathways relating to energy reserve substances (oils, sugars) and lignocellulosic cell walls, as well as ways of measuring the feedback effects of the major physiological plant functions of growth and development. Target species are first and foremost those which are already cultivated and well understood. In temperate countries, they include cereals (wheat, maize), oilseeds (rapeseed, sunflower, soybean, flax) and ligneous plants (poplar, locust, pine, eucalyptus). In tropical countries, they are chiefly sugarcane, sorghum, maize, oil palm and cotton. Other species are candidates for biofuel production, and include jatropha, Miscanthus, jojoba, cuphea and crambe. Finding other possible plants that could produce substances of interest is still a task to be accomplished. This requires that a systematic study of candidate plants be carried out and an international structure of biological resources created, while respecting living organisms and the advantages of biological resources. The range of possible plants becomes even greater when new species are considered, from micro-organisms (microalgae, cyanobacteria) to eukaryotes, both in open systems and in photobioreactors. It can be hoped that an integrated approach to biomass use, as conceived in the biorefinery concept, can reconcile food and chemical uses and optimise the global energy balance. As such, this approach should allow new agricultural areas that are today unsuitable for cultivation to be exploited for existing food and non-food uses that are well understood.

Microalgae

Single-cell microalgae (cyanobacteria and eukaryotes) are able to photosynthesise, which allows them to store energy, including fats and starches. Today, several thousand tonnes of microalgae are already being produced for products with high added value, such as food for humans or animals. Spirulina, for instance, is used to obtain beta-carotene. Microalgae that store fat are currently of the most interest for bioenergy production. The fat molecules are recovered via centrifugation or filtration, then extracted using hexane, and processed into biodiesel or biokerosene. Productivity calculations of small-scale tests have shown better yields in solar energy conversion from biomass than from terrestrial plants (up to 4 % compared to 1 %). This high potential productivity is one of the reasons for the enthusiasm for microalgae. However, this optimism must be put into perspective by emphasising that the comparisons (microalgae versus terrestrial biomass) have been made between systems that are at very different stages of maturity, and on very different scales (in the laboratory versus on an industrial scale).

Cultivation in a saline aquatic environment abides by the same principles as that of terrestrial plants, with the only concerns being to provide CO_2 and "directing" the microalgae's metabolism via stress to store fat. A mere 20 species of microalgae are being studied today, while several thousand exist naturally. With regards to the difference of scale, microalgae can be cultivated in photobioreactors, raceway ponds or open ponds, with culture densities ranging from 1 to 10 g/l. The use of open ponds for microalgae cultivation requires large water surfaces near coastal areas that do not compete with agricultural land (see Chap. 7). These crops are of particular interest when coupled with methanation, since they allow nutrients (with the exception of carbon) to be recycled, hence creating a circular bioeconomy.

How These Prospects Change the Hunger Equation

These new potential biomass uses will increase in value and most likely strengthen the role of agricultural production in world economies. With regards to world hunger, biomass uses will come into play at two levels. First, in the competition between food and non-food land uses—a phenomenon that is already underway and which will be discussed further in Chap. 7—and second, in terms of opportunities for the different agricultural sectors throughout the world that have land and favourable conditions for producing biomass, both in developed and developing countries. Beyond the different technical and economic issues, it is of fundamental importance to consider the impacts that such changes will have on world agriculture, taking into account both the opportunities created and inevitable drawbacks. It is absolutely essential to anticipate necessary worldwide regulations, so that the world's poorest countries and their inhabitants do not find themselves yet again in a losing position. This means making sure that competition for biomass does not exacerbate undernourishment and hunger.

Competing with Food Production

The prospects discussed above lead to the emergence of new, considerable demand for agricultural commodities for energy and chemistry purposes. Demand must be effective, at least in the long run and the first challenge is land use competition. "Land grabbing" has already been reported in some developing countries, and is founded on the desire to grow energy crops (see Chap. 7). This has been particularly true for first-generation biofuels. When the crops in question are also used for food (cereals, oilseeds or sugar plants), the market will respond to massive demands for energy uses, and even more so if they are encouraged by strong supportive policies. There will be tensions, if not about quantities, then certainly regarding prices; the impact on the food situation for poor populations, farmers and consumers must be analysed and well understood, unless the world is prepared to face an increased risk of hunger riots. Second-generation biofuels should lead to decreased risk of competition with food production. This will be true when they are produced using residues and waste [14], in addition to wood obtained through thinning operations in regenerated forest stands. This will not be the case as long as raw materials come from dedicated crops, and in particular, annual food crops used in their entirety. This issue is discussed further in Chap. 7, which deals specifically with available land area for crops and meadows, and the competition between alternative land uses.

The viability of family farms throughout the world is already in jeopardy today, which can be felt on economic, social and environmental levels. The structure of agricultural production, especially in developing countries, is fragile and leads to disadvantages and distorted views. New buyers entering the market are likely to upset the positions of current players and lead to changes that will not occur naturally or spontaneously, at the expense of the producers. Furthermore, because of the industrial approach to biomass processing, agrarian systems will be pushed into business-like structures rather than remaining family farms, which will not have immediate comparative advantages. The increasing fragility and exclusion of family farms—essential players in biomass development—may provoke serious consequences which must be planned for by policy decision-makers.

New Agricultural Opportunities for Developing Countries

Developing a profitable market for agricultural products or new forms of biomass could, however, have a very beneficial impact on agricultural and rural development, as well as for the people living in those areas. Such opportunities could allow producers, even small ones, a certain freedom of choice in putting their production on the market, if new sectors are created and sufficient budgetary resources provided to access these markets. The diversity of uses for a single product can be an advantage, creating flexibility for producers to optimise their revenue, even if this flexibility concerns mainly farmers having reached a sufficient level of technical knowledge and scale. This diversity is represented on a national scale by sugarcane policy in Brazil. Brazil's policy allows some flexibility with regards to cane processing, either into sugar or anhydrous alcohol to be added to gasoline, depending on the market. This flexibility is only apparent, however, if producers find themselves in a contractual obligation to deliver their harvest at the contracted price and unable to obtain a higher price by selling on the food market. But contracting can also be beneficial if market prices collapse. Moreover, bioenergy and green chemistry technologies give hope to the possibility of average-sized and decentralised production units, reducing logistical and transport issues. As far as energy is concerned, the ability to produce engine fuel locally, or electricity without being part of a national network, would be highly beneficial to local development.

All the elements needed to outline a clear plan related to new bioenergies and green chemistry for developing countries are not yet available. Nevertheless, some comparative advantages could be favourable to tropical regions, and in particular those that have the right climate conditions in terms of sunshine and precipitation. A "speciality" green chemistry, which would require labour-intensive agroecology, could provide opportunities for family farms, e.g. for cashew oil or spice farming. It could also apply to domesticating and producing new species from adaptation regions which have increased functional or chemical characteristics. What is beneficial for agriculture is not necessarily so for the struggle against undernourishment or hunger. Rather, favourable impacts are seen in the development of rural areas where populations suffering from hunger live (see Chap. 9). It is important to determine under what conditions—technological, organisational, economic and political—these opportunities help agricultural producers, increase their revenue and reduce the risks that weaken them. The cognitive and technological challenges of biomass non-food uses should be underscored, as well as the role that public policy plays in this area.

Cognitive and Technological Challenges

Whether in developed or developing countries, new biomass uses and the potential conflicts they may create for food bring up many issues. Today, energy security is as much a strategic concern as a food security issue. As for food, each country must set out to find a suitable energy mix to meet its needs. The possibilities for biomass use, linked to the development of other energies, whether renewable (wind, photovoltaic) or not, can change the current situation characterised by an energy supply that is dominated by a small number of oil and natural gas producing countries. The opportunity for change also depends on support for research.

White and green biotechnologies are major driving forces in the adaptation and processing of biomass, from enzymes to synthetic biology. Herein lies the major cognitive and technological challenges to examine in order to understand and represent the functions of living organisms and their components, as well as how they interact with human activities. Existing biotechnical and socio-economic knowledge must also be integrated, on various scales of time and space, to find solutions on a scope large enough to be applied to energy and chemical systems, while still appropriate for food systems. By bringing all of this knowledge together, new methods and forecasting tools can be developed to help private and public decision-makers. Finally, turning research into applicable—and, of course, sustainable—technology, equipment, processes and products can only benefit from joining comparable efforts being made concerning food systems. In fact, the problems are one and the same. If both areas can compete, then both can be complementary. These far-reaching changes, which would have an impact on agricultural activity, come with the assumption that the regulatory means allowing the greatest number of players to seize the food and non-food production opportunities available be carefully analysed.

The Role of Public Policy

In such a context, public policy clearly has its role. Public policy is a determining factor in the current expansion of bioenergy and bioproducts, whether it consists of specific incentive policies or regulatory measures that take the form of standards (e.g., the mandatory incorporation of a certain percentage of biofuels), fiscal advantages or certification measures. Adopted by a growing number of countries, such policies set objectives tied to the strategic and environmental concerns discussed previously: reducing oil dependency, finding oil alternatives, and reducing GHG emissions. They also look to support the revenues of agricultural producers. The desire to create new agricultural markets and agro-industrial sectors was decisive, especially in the United States and in several European Union countries. Finally—and fortunately—policies favouring the development of bioenergy and bioproducts include certain conditions regarding the sustainability of promoted renewable energies and take into account, albeit in a somewhat limited manner, the issue of land use conversion.

These policies can be considered an anticipatory support to a burgeoning industry. Both standards and certification measures provide incentive to change. In the European Union, for example, implementing directives relating to reducing volatile organic compounds, or the REACH regulation, has encouraged numerous sectors to research or anticipate solutions to replace petrochemical products. In general, regulations are a way to influence, whether by positive or negative reinforcement, bioproduct sectors. Measures that encourage and regulate ecodesign, biodegradability standards, thermal regulations, diverse regulations on health or air quality, regulations on packaging and detergents, and many others affect the choices made by operators. Certification measures are also important, especially those concerning environmental certification such as the EU energy label "Ecolabel". Considering the growing demand for environmental and health information on products, the qualification of bioproducts can be an important driving factor. Life Cycle Assessments and sustainability criteria applied on a large scale thus need to be comprehensible, transparent and reliable, because their potential influence on the market is considerable. The only thing left, then, with regards to bioenergies, and in particular, first- and second-generation biofuels, is the thorny issue of land use competition. This issue will be discussed in more detail and in a broader fashion in Chap. 7.

References

1. Wirsenius S (2008) Global use of agricultural biomass for food and non-food purposes: current situation and future outlook. Working Paper, Chalmers University of Technology, Department of Energy and Environment, Gothenburg, Sweden. See also: Wirsenius S (2003) The biomass metabolism of the food system: a model-based survey of the global and regional turnover of food biomass. J Ind Ecol 7(1):47–80
2. Jancovici JM, Grandjean JM (2006) Le plein s'il vous plaît! La solution au problème de l'énergie. Seuil, Paris, 185 p
3. Dronne Y, Forslund A, Guyomard H (2011) Les biocarburants de deuxième génération et la compétition pour l'usage des terres. OCL (Oléagineux Corps gras Lipides) 18(1):1–9
4. Cormeau J, Ghosse G (2008) Les biocarburants de deuxième génération: semer aujourd'hui les biocarburants de demain. In Déméter 2008, Économie et stratégies agricoles, p 225–302
5. MEA—Millennium Ecosystem Assessment (2005) Ecosystems and human well-being: synthesis. Island Press, Washington D.C., 155 p

6. Searchinger T, Heimlich R, Houghton R et al (2008) Use of US croplands for biofuels increases greenhouse gases through emissions for land-use change. Science 319:1238–1240
7. Fargione J, Hill J, Tilman D, Polasky S, Hawthorne P (2008) Land learing and the biofuel carbon debt. Science 319:1235–1238
8. Dorin B, Gitz V (2008) Ecobilans des biocarburants: une revue des controverses. Natures Sciences Sociétés 16:337–347
9. Gohin A (2008) Impact of the European biofuel policy on the farm sector: a general equilibrium analysis. Rev Agric Econ 30(4):623–631
10. Bureau JC, Guyomard H, Jacquet F, Treguer D (2010) European biofuel policy: how far will public support go? In: Khanna M, Scheffran J, Zilberman D (eds) Handbook of bioenergy economics and policy. Springer, pp 401–425
11. Anastas PT, Warner JC (1998) Green chemistry: theory and practice. Oxford University Press, Oxford, 135 p
12. Ademe (2008) Référentiel méthodologique pour la réalisation d'analyses de cycle de vie appliquées aux biocarburants de première génération en France—Synthèse
13. WTW EUCAR/JRC/CONCAWE (2006) Well-to-wheels analysis of future automotive fuels and power trains in the European context. http://ies.jrc.ec.europa.eu/wtw.html
14. Blascheck H, Ezeji TC, Scheffran J (2010) Biofuels from agricultural wastes and byproducts. Wiley-Blackwell, 262 p

Chapter 6
Managing Ecosystem Services

Nature provides men and women with—or makes it possible for them to produce—essential goods and services. These goods are not just the result of simple extraction, as with fossil fuels, but rather are the result of numerous ecological processes that depend on characteristics of the environment that are sensitive to human activities. Nature supplies the foundation and walls that hold up the history and organisation of our societies. Since the UN's launch of the Millennium Ecosystem Assessment (MA) in 2001, the issue of goods and services provided by nature has taken on increased importance. The concept is now known as "ecosystem services", although the use of the word "services" can be somewhat misleading, because it also includes agricultural production. The MA distinguishes four major ecosystem service families:

- Supporting services, which ensure the basic ecosystem functions, such as the carbon cycle;
- Provisioning services, which include food, water and genetic resources;
- Regulating services, from limiting flood risks by plants to plant pollination; and
- Cultural services, including the heritage value of a landscape or protected species.

These four service families are shown in Fig. 6.1.

Many natural resources are limited, and the functional processes at work in the biosphere are fragile. This observation is not new, and is increasingly documented and accepted. Crops and livestock farming occupy a major part of the Earth's land area. As such, agriculture is often singled out. Means of counterbalancing the negative environmental impacts of agricultural production have long been the subject of research and technical, economic, legal and political developments. The concept of ecosystem services has become a mainstream idea and is no longer confined to the fields of science and technology. That first approach, focused on agriculture and its impacts, has been expanded, and agricultural production is now understood to be only one part of the many services mankind receives from ecosystems, which affect food, lifestyles, health and well-being. These two approaches were once considered separate and related to different schools of thought. Today, their paths have come together, and are in fact complementary, first because agriculture holds a considerable

M. Guillou, G. Matheron, *The World's Challenge*,
DOI 10.1007/978-94-017-8569-3_6, © Éditions Quæ, 2014

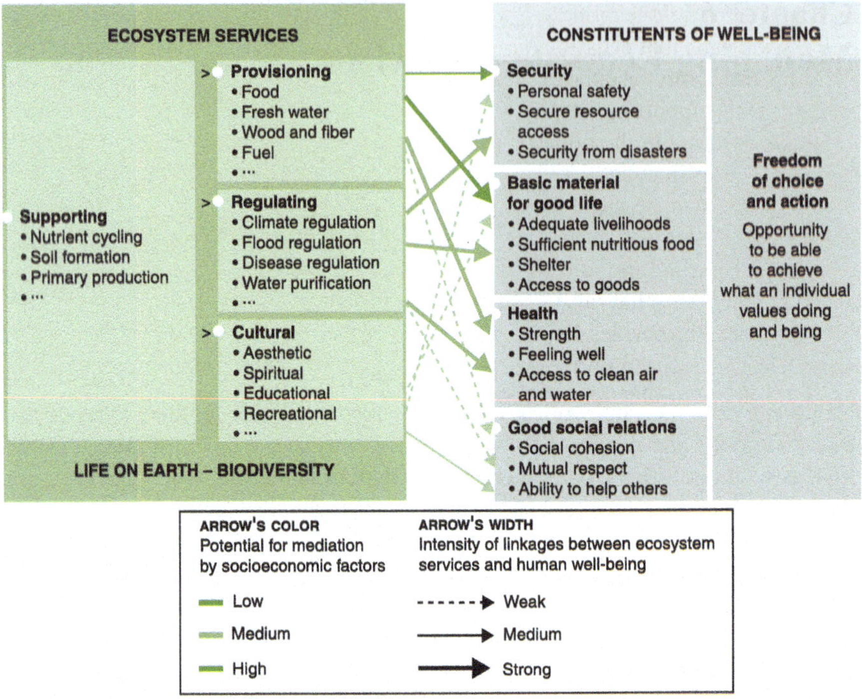

Fig. 6.1 Classification of ecosystem services and linkages with human well-being (MA 2005). (Reproduced with the permission of F. Bourin Publishers)

place in ecosystem services—whether producing or using them—and second, because agricultural strategies can no longer be established without considering their interactions within the entire structure of ecosystem services.

The MA report was published in 2005 [1]. The concept of ecosystem services quickly spread in the political sphere because it met both the expectations and the needs of public policy makers [2]. Within the research community, however, the concept had gradually been gaining ground since the 1970s through the work of the Club of Rome, and later through more precise and analytic work, such as in the article by Costanza et al., *The Value of the World's Ecosystem Services and Natural Capital*, published in 1997 [3]. It is important to distinguish between two notions that are sometimes mistaken to be the same: ecosystem services and environmental services. Environmental services are actually a part of ecosystem services. In other words, ecosystem services encompass much more than the simple environmental services provided by ecosystems. Ecosystem services are an interpretation of what people get out of ecosystems. It is fitting, then, that the economic, social and human sciences define the boundaries of such services. Biotechnical or biological sciences, particularly agronomy and ecology, are even more applicable to the processes that result in these services. This duality is not always clearly understood and compounds the lack of understanding resulting from the disconnected history of the relevant scientific disciplines. On one hand, the interconnectedness of ecological processes means that ecosystem services are closely related to each other, whether in terms of synergy or oppositions. This interconnectedness creates a top-level hierarchy, with supporting and regulating services determining the provisioning and cultural services. On the other hand, the expectations of the different players with regards to ecosystem services results in other possible, less-clearly defined hierarchies: for instance, people must be fed first, then biodiversity can be protected, then land can be used for cities, and so on. These "social" hierarchies are not set in stone; they vary according to the relevant ecosystem, history, culture, stage of economic development, etc. They also vary between developed and developing countries. Current research, besides linking the many relevant sciences to pluridisciplinary approaches, is focused on the two hierarchic viewpoints outlined above. Practices and agricultural production systems that are more respectful of ecosystem functions must be developed for each of the pertinent environmental services, and integrated management strategies must be devised that take into account societal priorities and local specificities.

Agriculture and Ecosystem Services

The relationships between agricultural activity and the state and functioning of ecosystems have been the subject of much research. In France, these relationships have been examined in collective scientific expert reports, which have brought together various scientists from different research organisations and use internationally available scientific publications for their research. The reports in question mainly concern developed countries. They pertain to subsets of ecosystem services, such as soil carbon storage, GHG management, and quantitative water management. Reducing pesticide use was also a subject of a collective scientific expert report, rapidly completed by another study, the Ecophyto R&D, focused on the possibilities of reducing pesticides in French agriculture and the consequences of that reduction. Research focusing specifically on ecosystem service families and wider ranges of ecological processes is more recent. In France, they were also examined in another collective scientific expert report dealing with the links between agriculture and biodiversity [4]. It was essentially thanks to "biodiversity" making its entrance that the continuity between agronomy, ecology and the economic, social and human sciences began to develop around ecosystem services.

The Hydrologic Regulating Service and Quantitative Water Management

Water is a determining component in agricultural production, whether that production is irrigated or not. Its use in agriculture also has a significant effect on the regulation of the water cycle, as well as on the provisioning of other ecosystem services. The hydrologic regulating service includes all environmental processes which affect the quantity and quality of both surface and underground water. The impact of a forest near a catchment area on water turbidity and river flow is an example. Similarly, agricultural activities can have a strong influence on the quantity and quality of water depending on their needs. Furthermore, climate changes, especially the increasing frequency of extreme weather events such as droughts and flooding, make the relationship between the hydrologic regulating service and other ecosystem services all the more apparent.

The collective scientific expert report "Drought and Agriculture", published in 2006, was limited to France and focused only on agriculture, with forests being excluded from the study [5]. The detailed analysis of regional and local geological, pedologic or climatic particularities proved to be indispensable, as the diversity of such specificities excluded any generalisation right from the start. The report explored the feasability of implementing agricultural production systems better suited to water shortages, as well as improved genetic and agronomic prospects that could be used for such a purpose. The report looked possible ways to reduce water use in agriculture under acute shortage conditions and heightened competition with other uses. Several promising possibilities were identified:

- in genetics, for species such as sorghum and sunflower, strategies aiming to avoid water stress or that seek to differentiate the growth of different organs during drought periods;
- in cropping or livestock farming systems, strategies using technical operations which optimise the quantity and rhythm of irrigation coupled with the diversification and spatial distribution of crops.

Such shifts imply coordinating the different players involved in a single area and adapting economic circuits, especially to further develop—or even create—sectors for species that would be introduced and that today do not have a significant presence on the market. They also require the participation of public authorities based on three principles: increasing water availability to better manage supply so that it meets demand; saving water so that demand can be adapted to supply; and finally, compensation after the fact by public or private insurances. Managing priorities must be coordinated on a local level so that short-term actions (irrigation bans, exceptions accorded on turning fallow land into pasture) can more towards structural actions pertaining to volume-negotiated water management on a physical, economic and political basis.

Using the environmental services concept has permitted localised actions to protect catchment basins through Payments for Environmental Services (PES) programmes. Under such schemes, land uses that have a positive effect on the hydrologic regulating service are rewarded. More precisely, the players benefiting from better water quality and/or who have easier access to greater quantities of water accept to pay those who, through their actions (chiefly farmers who change their practices and systems), bring about this qualitative and quantitative water improvement. This remuneration gives the second category of players the chance to move forward by compensating—or even overcompensating—the costs involved in changing practices, systems and techniques. Steps have been taken in line with these principles in many developed countries (in Vittel, France; Munich, Germany; Chesapeake Bay, United States) and developing countries alike (Bolivia, Costa Rica, Ecuador). PES is a promising system that should be researched further, both for the hydrologic regulating service and for other environmental services. It has the advantage of overcoming budgetary shortfalls at national or local levels; however, there are also certain risks, such as inefficiency or corruption, which must not be disregarded.

Reducing Pesticide Use

Pesticides refer to three product families that aim to protect crops from weeds (herbicides), fungi (fungicides), and insects (insecticides). The non-desirable direct effects of pesticides are their toxicity to humans and biodiversity. Considerable research in ecotoxicology means these effects are generally well known, except—and not incidentally—for the effects of long-term exposure to low amounts of pesticides and the effects of combining certain products. Indirect effects on ecosystems are more poorly understood because they are felt through the functional processes of ecosystem services. Because of the lack of precise knowledge about alternative techniques, their costs and impacts, as well as the absence of prioritisation of possible economic, social and political compromises, the general tendency is to simply reduce the quantities of pesticides used.

In France, this tendency has led to a commitment, made during the French Environment Round Table (*Grenelle de l'environnement*), to reduce pesticide use by 50 % by 2018 "if possible". The study carried out prior to the Round Table by the French National Institute for Agricultural Research (INRA), commissioned by the Ministers for Agriculture and Environment [6], made it possible to outline scenarios for reducing pesticide use for the main groups of agricultural production in France (arable crops, fruit orchards, vegetables and viticulture). The experts looked at known production management strategies, classed according to five levels: intensive protection; supervised protection; integrated protection; integrated production, particularly with regards to crop rotation; and organic production. The study evaluated the consequences of applying these five strategies across France with regards to pesticide use, agricultural yield and farmers' revenues.

Overall, with all the necessary reserves taken due to the nature of methodological simplifications in this type of study, it seems that the generalisation of integrated crop protection practices reduces pesticide use by 20–30 % depending on the crops. Most often there is no loss in revenue and around 6–8 % lower yields for cereals and oilseeds, with greater decreases for other crops such as certain fruits. To reach a 50 % reduction in pesticide use, not only will production techniques need to undergo profound changes, but also those for crop rotation, which should follow recommendations set out by integrated production and organic farming. But under these conditions, revenues and yield will be markedly different, and the various production sectors will be seriously affected. In short, the resulting disruptions will have a significant impact, whether beneficial or destabilising, on the landscapes and ecological processes that determine ecosystem services, as well as on the activities and social organisation of the land. In addition to its specific objective, the study enabled more general observations to be made about land use issues, including land use conversion and use intensity. Modifying agricultural production strategies in ecological areas with low to very low yields can quickly lead to significantly higher yields. Moderate use of pesticides and the development of protective crop practices based on natural processes, such as biological control, combined with the conservation of diverse agricultural productions and landscapes, can most likely increase agricultural product volumes without serious disturbance of ecosystem services. However, carelessly implementing a model for maximum yield could destabilise ecosystem functions.

Myths and Truths About Fertilisers, Livestock Farming and GHGs

Storing carbon from organic matter in the soil helps limit the amount circulating as GHGs in the atmosphere, thereby contributing to the environmental service of climate regulation. Soil carbon stocks are significant: they equal double the amount of carbon in the atmosphere. INRA's collective scientific expert report on increasing carbon stocks in agricultural soils in France demonstrated that changing soil use and agricultural practices led to considerable increases in organic carbon stocks in the soils examined [7]. According to the different scenarios, storage potential could reach 1–5 million additional tonnes of carbon each year over 20 years. These figures, even if they only represent a small percentage of France's total GHG emissions, cannot be ignored. That said, the expert report does temper the tendency to be overenthusiastic. It is important to note that increased soil carbon storage only has a temporary effect; benefits cease as soon as the maximum storage capacity of the soil is reached (a timeline of about 20 years). Also, in this case, increased carbon storage in soils implies radical land use changes, with an increase in forest surface areas and the conversion of cultivated land into permanent grasslands. That said, the report noted a more positive development, with possible synergic effects from the increase in carbon stocks in the soil on other ecosystem services, such as erosion control, soil quality, water quality and biodiversity conservation. The role agriculture plays in generating other GHGs is the subject of numerous studies and is a controversial topic. Methane from animal farming and nitrous oxide from fertiliser are receiving particular attention. According to some estimates, agriculture contributes to between 25 and 75 % of total emissions of nitrous oxide, a gas with a warming potential 300 times greater than CO_2. These data, often collected in developed countries, sometimes lead to categorical recommendations to reduce fertiliser use, which would result in decreased agricultural production.

Such partial analyses should be taken with a grain of salt, and are best replaced by a more far-reaching vision that takes land use issues into account. Glendining et al. [8] suggest that agricultural production systems currently used in the United Kingdom have a lesser environmental impact than more extensive systems when assessed using GHGs and biodiversity conservation indicators. While the more widespread systems are certainly more economical in terms of outside input (mineral fertilisers, crop treatment products and livestock feed), they are more demanding of the land. Thus, a scenario of significant expansion of cultivated land to meet the world's increased food needs by 2050 could in fact have a more negative overall effect on climate regulation by considerably increasing total GHG emissions. In the end, this type of scenario would be worse, at least in terms of GHG emissions, than more intensive agricultural practices on existing cultivated land, which would greatly increase yields in areas where they are still relatively low [9]. In summary, the environmental gains of using fewer mineral fertilisers must be balanced with the effects of upsetting the ecosystem services hierarchy due to the necessity of converting new land to crop use [10].

Soil Properties and Ecosystem Services

The notion of soil fertility does not have a precise scientific meaning. It can be defined as the capacity of a soil to produce a crop, sustainably and abundantly. However, soil fertility is inextricably linked to climatic conditions, which affect production potential, and human activities. The detailed analysis of the role soil plays shows that it holds a central place for the services it provides and processes it shelters. The recent inventory of several studies by Dominati et al. [11] demonstrates that it is possible to clearly classify the services performed by soils and the mechanisms that enable such services. Soil, a physical support which is favourable to root development and nutrient absorption, also helps limit the effects of erosion, regulates water circulation and controls water pollution, recycles organic matter, regulates GHGs, and controls diseases and pests. Numerous biodiversity agents in the soil contribute to these services, from micro-organisms that transform litter to those involved in the interactions with roots. There are a considerable number of symbiotic processes that are essential to plant growth, including organisms that control diseases and others called "ecosystem engineers" that determine soil structure [12]. As is always the case when dealing with ecosystem services, there are many interconnected effects. If soil functions determine the yield of the crop grown in it, the soil is also altered by that crop. Different studies mentioned in the collective scientific expert report "Agriculture and Biodiversity" show a positive effect of functional plant diversity on soil properties. For example, pasture legumes and their introduction in crop rotations positively affect nitrogen availability in the soil. Furthermore, the texture and depth of the soil play a major role in hydrologic regulation by adjusting flow rates.

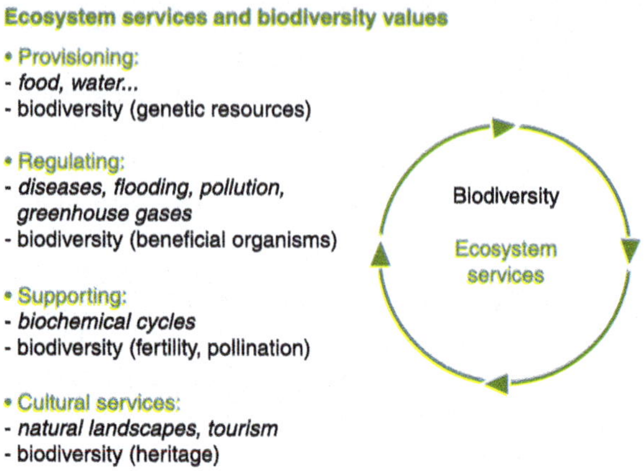

Fig. 6.2 Examples of interactions between biodiversity and ecosystem services. How genetic diversity influences the ability of an ecosystem to provide food. (Reproduced with the permission of F. Bourin Publishers)

Biodiversity's Key Role in Ecological Regulating Services

It is not easy to take into account—and all at the same time—the different ecosystem services, ecological process families that regulate them and the economic, social and political strategies adopted to manage them. There needs to be a simple and symbolic approach that takes into consideration ecosystem services. Biodiversity held the role of catalyst, and it still plays that role today, illustrating the relationship between ecological processes and ecosystem services [13].

The Many Faces of Biodiversity

Biodiversity refers to all life forms on Earth, "the living fabric of our world", as Robert Barbault calls it [14]. The term biodiversity was coined in 1985 by Walter G. Rosen during preparations for the National Forum on Biological Diversity, held in 1986 and organised by the National Research Council. It encompasses a wide range of complex subjects, because it deals not only with the diversity of living species present within an ecosystem, but with populations belonging to the same species or an individual's genes within the same population as well. Biodiversity is also defined through the study of the dynamic relationships that tie living beings together in the different ecosystems, and the many interactions and evolutions between these ecosystems on the Earth's surface. A common good and responsibility, biodiversity is a factor in most ecosystem services, as shown in Fig. 6.2. Activities in the agricultural and forestry sectors are at the heart of these processes. In addition to remarkable biodiversity, exemplified by symbolic species such as large terrestrial and sea mammals, "ordinary" biodiversity is essential in agricultural and forestry practices.

Agriculture Uses Biodiversity: The Importance of Genetic Resources

History has shaped agriculture by focusing on and managing the transformation of solar energy into plant matter, which can be used by mankind both for food and non-food uses, and producing animal-derived goods (meat, milk, eggs) from that plant matter. The species, varieties and breeds, be they cultivated or raised, all come from genetic resources that are initially available in nature. The first characteristic that links agriculture and biodiversity is that agricultural production is, by definition, a use of biodiversity. In the 1960s, research on genetic resources was mainly focused on productivity per land unit and innovations that would enable moving beyond the factors that limited production under standardised conditions. Differences between environments and situations were eliminated to the fullest extent possible by production techniques and using external input (irrigation, fertilisers, pesticides, etc.). As was to be expected, very few species and varieties proved productive and efficient under these conditions. Because of a strong consistency in production systems and in the link between variety selection, crop management and the processing sector, the approach led to significantly increased yields and the ability to deal with hazards—the cost too often being environmental damage from fertiliser and pesticide use that exceed the absorption capacities of the ecosystems cultivated under such conditions.

But we no longer live in the same world: climate change, increasingly scarce fossil fuels, damage to the environment and natural resources, diversified market and consumer demands all clamour for a shift—if not a revolution—in agricultural practices and production systems, which must be less energy-hungry, more respectful of natural resources and the environment, and more adapted to local particularities without trying to simply eliminate them. Chapter 8 is devoted specifically to the issue of needing renewed practices and agricultural systems. This chapter will focus on using genetic resources to the fullest, which must play a role in this agricultural renewal, one which is adapted to a wide range of contexts and objectives. It is a veritable paradigm shift: genetic resources must not be viewed simply as a way to produce more; they are the key to integrating agriculture into the overall functioning of ecosystems.

Several research programmes have set this goal; they bring together many areas (agronomy, ecology, economic and social sciences, among others), and are often drawn up within a partnership framework between professional agricultural organisations. Furthermore, "domesticated" biodiversity is now considered a stakeholder in general biodiversity. In France, the Bureau des ressources génétiques merged with the Institut français de la biodiversité to create the Foundation for Research on Biodiversity (FRB) [15]. At the heart of its Steering Committee, associations for the protection of nature work alongside industry, seed companies and rural managers. The FRB, which is responsible for France's national strategy on biodiversity research, manages national and international research programmes to classify, maintain and disseminate "ordinary" genetic diversity.

Agriculture and Biodiversity: Moving from the Plot to the Landscape [4]

As was the case for agronomy, for a long time the study of agriculture's effects on biodiversity was centred solely on the scale of a plot and the agricultural practices conducted on it. However—and as was shown in the "Agriculture and Biodiversity" expert report—biodiversity is also present in many processes that are relevant on a much larger scale, now referred to as a "landscape". In particular, the report showed that in temperate zones, agriculture's impact on biodiversity depends on three concurrent characteristics of the landscape in question: the degree of agricultural production intensity (as in production per land unit); the degree of uniformity in production (in terms of cultivated species and varieties); and the degree of elimination of permanent grasslands or semi-natural areas such as field margins, hedgerows and tree groves. This third parameter plays a very limiting role: beyond a certain proportion of grasslands and semi-natural areas, lowering the degree of intensity or re-diversifying crops is not enough to correct for losses in biodiversity. During the Environment Round Table, the idea of green infrastructure was introduced, which basically aims to restore the spatial continuity necessary to maintain and allow wild populations to migrate. For a wide range of organisms, it is in the structure of various landscapes as a whole—and not simply in how they are interconnected—that the major biodiversity regulating services are found [16].

"Behind the Scenes": The Functional Trait Approach

The complicated processes that rely on biodiversity participation are rivalled only by the complexity of the related ecosystem services. Pollination or controlling pests using natural predators implies mobile organisms that are sensitive not only to the surrounding spatial structure, but on the scales of the plant, plot and landscape. But the structural stability of the soil, the conservation of its fertility, and water availability for crops are ensured by organisms that are not very mobile, such as fauna and soil micro-organisms, which are more dependent upon the direct effects of plot management. For its condition to be perceptible both to the general public and policy makers, biodiversity needs a functional indicator comparable to the "tonnes of carbon" used in GHG management. After the importance of moving to a landscape scale, the second major conclusion of the "Agriculture and Biodiversity" report was that it is just as important to consider biodiversity functioning, its dynamic and its evolution—and how it is affected by global changes and human activity—as it is to look at its condition in a given time and place. To develop strategies that can be understood by the public and that are usable by policy makers, tools need to be available that illustrate this functioning. Here again, research progress is promising.

Ecosystem services are interconnected by functional traits, as shown in Fig. 6.3 [17]. These traits are the associations of morphological or physiological

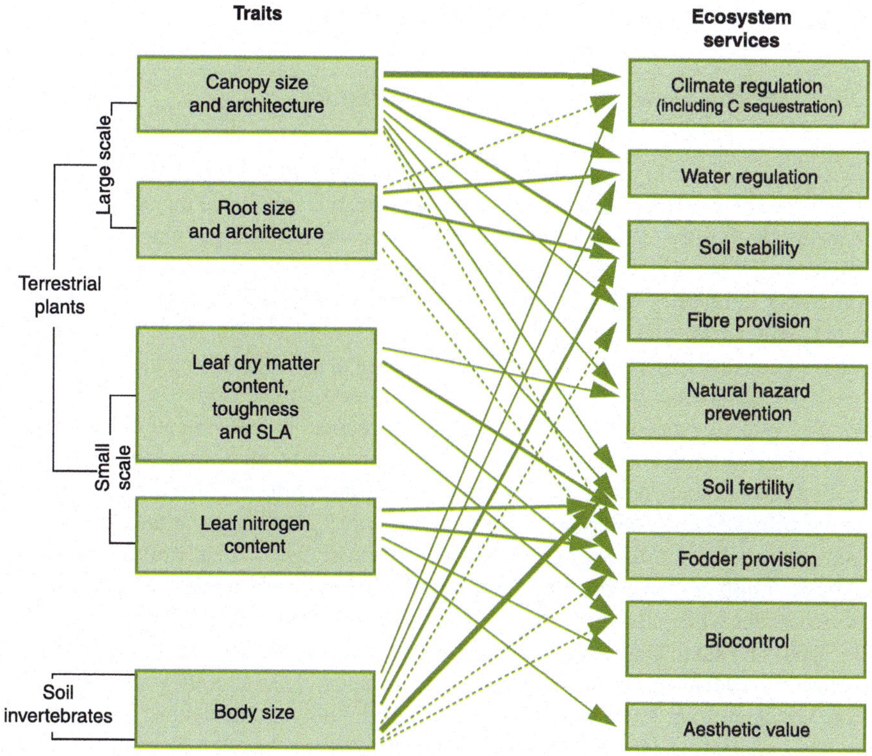

Fig. 6.3 Measurable morphological and physiological traits give information on services provided by ecosystems. (Source: [17], reproduced with the permission of F. Bourin Publishers)

characteristics by which living beings are able to grow, reproduce, defend themselves, etc. They have the advantage of being more easily measurable than the functions themselves. They are then compared to the characteristics of the environment. For example, functional traits of grassland plants are linked to the characteristics of micro-organism communities in the soil, to the particularities of nutrient cycles, and grassland exploitation methods by herbivores. A profile of functional traits is then established from a set of biodiversity characteristics, which means the number of parameters measured in the field can be considerably reduced. Moreover, these functional trait associations are even more representative of the way ecosystems function than are lists of species, and can be linked to actual or potential ecosystem services. These are huge accomplishments—even if there is still progress to be made in order to become completely operational—towards developing usable indicators for policy makers and creating forecasting tools that are more efficient when faced with uncertainty [18].

Using and Managing Ecosystem Services

The Starting Point for an Integrated Vision

Decrypting the ecological processes that enable agricultural production and condition ecosystem services is a crucial point for research and for taking it further. Held back by the history of their disciplines, agronomists and ecologists have had to take a path based on aggregates, using elementary functions to move towards integrated processes in order to build a consistent theoretical corpus and tools suitable to field work. Right from the start, economists and sociologists were faced with a global view of ecosystem services, even if they were not always known by this name. How were they to go about collectively evaluating these services? How could they come up with effective means to manage them, while reconciling the potentially contradictory interests within a society—farmers, urban planners, environmental activists—with those of other societies having varied histories, cultures and different immediate priorities? How could they determine the priority directions to take for a given ecosystem's future? How could they integrate those with global priorities, such as feeding 9 billion men and women on a sustainable basis while preserving ecosystem services, but while also respecting local situations when developing strategies?

This is no small challenge. The most logical place to start is with the idea of evaluating ecosystem services from a monetary perspective so that they are fully taken into consideration when making individual and collective decisions. The working method consists in identifying the services, measuring them, and assigning them a value. This plan is not without its difficulties. From a scientific viewpoint, the three stages of identifying, measuring and assigning monetary value are issues that have yet to be resolved. And from a societal, moral and even ethical standpoint, there are many who refuse to place a price on living organisms.

Economic, Legal and Political Evaluation Aspects

Assigning a Value to Environmental Services and Biodiversity

Bringing together so many different services as agricultural production, the heritage value of biodiversity and its effects on tourism, the role pollinators play in plant production or that of micro-organisms in soil fertility under one economic framework implies having a benchmark that makes comparing apples to oranges possible. Economists have one solution, albeit an imperfect one: attributing a monetary value to the different services, estimated using methodologies that have been adapted to each service. Thus, provisioning services are assessed by their market values; regulating services by opportunity costs of their degradation and restoration; and cultural services by contingent valuation, hedonic pricing, and revealed preference methods. For biodiversity, depending on the roles it plays in the ecosystem, the same methods used to assign values to regulating and cultural services are applied.

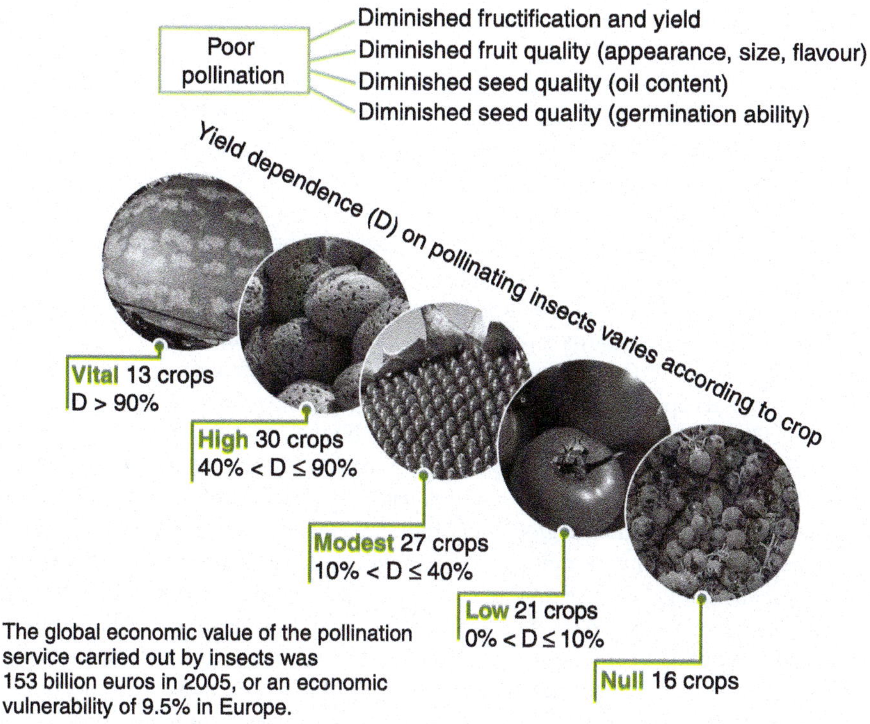

Fig. 6.4 The example of pollination. (Reproduced with the permission of F. Bourin Publishers)

The sum of the values equals the Total Economic Value (TEV) of the environmental services and biodiversity. This TEV naturally depends on current knowledge of such services and biodiversity; it is not fixed in stone and varies according to the ecological, historical, economic and social context. Without dissimulating the disadvantages previously mentioned, it would seem that the advantages win out. Using this method means a single indicator can be used to summarise all available information. Making decisions is thus possible. When it works, the results can be spectacular. The Mexican government, for example, increased taxes on water to account for the ecosystem role of forests in catchment areas; total annual revenues collected went from 15 to 150 million US $ [19].

Insert 6.1. The Value of Pollination

A recent study by several European teams placed a worldwide value of 153 billion € per year on the ecosystem service of pollination. This value was determined by establishing the importance of pollination for different crops, then estimating the consequences of a reduction in or the complete disappearance of pollination of those crops, both quantitatively and qualitatively [20] (Fig. 6.4).

The "Polluter Pays" Principle and the Compensation Mechanism

Moving beyond economic assessments that have not yet been followed by sufficient concrete measures, policies that take into account environmental services and biodiversity can generally be lumped under one approach, as long as doing so remains socially and economically viable: the polluter pays principle and the compensation mechanism. A quick description of these two tools is in order, based on conclusions in a recent report (2009) by the French Strategic Council for Sustainable Agriculture and Agro-industry (CSAAD) [21]. The polluter pays principle is based on true cost accounting and integrating these costs into the resulting damages (and benefits, if there are any). As regards environmental services and biodiversity, the problem lies again with identifying, measuring and assessing the damage.

When (or if) applied, the polluter pays principle rarely fully covers the damage caused. The compensation mechanism is based on the idea that shares for environmental services or biodiversity are sold on a market to achieve a goal of "No Net Loss" (NNL). There are very few environmental service markets based on this principle. Examples include the global and regional (European Union) carbon sharing schemes and one for hydrologic regulating services in the United States. This is also known as the mitigation principle, according to which contamination or degradation of an ecosystem in a given location can be authorised if made up for in another location. When applied to agriculture, such a mechanism implies that environmentally- and biodiversity-friendly agricultural activities (and therefore those resulting in services other than just agricultural production) can issue shares for these services and biodiversity which are then acquired by legal entities whose activities degrade those same services and biodiversity. The exchange instruments can vary, with a natural vehicle being Payments for Environmental Services (PSE), which have already been discussed. In this case, the provider of a given environmental service sells said service to a buyer who can degrade that service in an equal amount. Aside from the difficulties in determining the zero sum equivalent, it is clear that the compensation mechanism is only able to maintain the status quo, unless overcompensation situations are created, such as making two shares equal to the right to degrade for one.

A Few Examples

With regards to French agriculture, the working method depends both on public actions and voluntary private initiatives. The Environment Round Table set many ambitious objectives: 20 % of land to be used for organic farming by 2020, pesticide use halved "if possible" by 2018, and an increase in farms with low energy dependence equalling 30 % in 5 years. Other measures aim to develop green infrastructure to promote ecological connectivity and to implement, within 5 years, a "high environmental value" certification for farms. Concerning ecological compensation, the initiative relies on the CDC Biodiversité, a subsidiary of France's Caisse des Dépôts et Consignations (CDC). The mechanism functions the same as was previously described: any loss in biodiversity in an ecosystem or habitat must be compensated in an equal amount in another location. However, the mechanism is only at its beginning; it is still somewhat fragmented and not structured to handle large areas over the long term. It is essentially aimed at land and infrastructure developers, while farmers are the ones in the position to provide environmental and biodiversity services, services for which they could also be compensated. The PES-based initiatives are most often used in catchment areas, and depend on the initiative of water companies. In the United Kingdom, a larger-scale experiment was launched in the form of an auction, allowing farmers to sell an environmental benefit, with the most attractive offers for the community being contracted.

In the European Union, current measures under the Common Agricultural Policy (CAP) focus more on limiting the negative impacts of agricultural production than on actually encouraging farmers to produce positive environmental externalities. This is especially the case with agri-environmental measures, which compensate potentially higher costs related to more environmentally sound practices, but which stop there, not rewarding additional efforts. In the United States, two main remuneration systems for environmental services produced by agriculture coexist. The first, already mentioned, concerns hydrologic regulation: it is built around services that farmers can sell based on a long-term partnership arrangement to restore certain ecological potentialities of land used for agriculture (wetlands, grazing land, etc.). The second is a trading scheme where shares can be bought for CO_2, water quality, biodiversity and urbanisation rights.

In developing countries, environmental protection measures and environmental service management are often carried out on a project-by-project basis. In many cases, these projects have proved to be quite effective. In North Vietnam, rather than constructing dykes, local communities have planted mangroves as protection against natural catastrophes that threaten 70% of coastal populations. The initial investment (1.1 million US $) was much lower than the annual cost to maintain the dykes (7.3 million US $). Similar flood control projects through the preservation of vegetation areas have been implemented in Brazil and India. Also in India, a project that was abandoned due to uncompensated destruction of ecosystem services should be noted. The project, in the Delhi region, planned to convert a 3,250-ha plain that was regularly flooded into a habitable area. This plain provided water, fodder, fish, recreational services and much more. Regulating the river's flow would have eliminated these services, estimated at 843 US dollars per hectare per year; this sizeable sum provided justification to abandon the project. In the same vein, a project to convert a mangrove into shrimp farms in Thailand did not come to fruition. The private net profit of the project was estimated at 1,220 US dollars per hectare per year, but the assessment did not take into account the products regularly collected from the mangrove (987 US dollars per hectare per year), costs of converting the site (9,318 US dollars per hectare after 5 years of use), and the cost of protecting the farms and coastline against storms (10,821 US $). It also ignored the positive effects of carbon sequestration by the mangrove[1].

[1] The examples given in this paragraph are taken from different TEEB reports, "The Economics of Ecosystems and Biodiversity", available from www.TEEBweb.org.

Going a Bit Further

Ultimately, public and private environmental protection measures are not yet quite up to task. This is true for environmental protection, and even more so for environmental services provision and management. From research to practical applications, there is still more progress to be made, despite the cornerstones of the foundation having been laid. The examples described above demonstrate the necessity to go further: building consistent and global measures that take into account environmental services and biodiversity within agricultural policies, and more generally within public policy. This imperative is clearly reflected in recent reports. Examples include the report by the French strategic analysis centre (CAS) on the economic approach to biodiversity and ecosystem services [21] and the international initiative, "The Economics of Ecosystems and Biodiversity", or TEEB study, hosted by the UN Environment Programme (UNEP) and financially supported by several countries.

The originality of the CAS report lies in its suggestion to create a consistent and global framework, from the classification of relationships between biodiversity and ecosystem services to their economic assessment and legal and political options. Refuting the argument that ecological processes related to biodiversity are too complex to be evaluated within a comprehensive scope, the report considers that the available knowledge is already sufficient to assess the state of biodiversity and its evolution over time in a given location, while remarkable biodiversity characteristics can be distinguished from ordinary biodiversity. The analytic method is strengthened by case studies, notably the example of temperate forests where taking into account the whole of ecosystem services significantly alters the vision. The major operational conclusion of the report is nevertheless that assigning a value to the different services of a given ecosystem is not enough to ensure that those services are managed well. There must be government intervention in order to create an institutional and regulatory framework to allow the market to develop; for numerous environmental services and biodiversity, there is simply no spontaneous private demand.

The optimum use of environmental services and biodiversity takes on another dimension when evaluated on a global scale. The different TEEB reports, which each time brought together more than 100 experts from some 40 countries under the coordination of economist Pavan Sukhdev, focused on analysing the relationships between the degradation of ecosystems—and notably the global biodiversity crisis—and persisting rural poverty. These researchers insist on the importance of making the value of natural capital visible and integrating this value into macroeconomic approaches. They also emphasise the significant need to make biotechnical knowledge, especially ecological, part of the economic models and note that the variety of priorities as regards services according to location make it difficult to extrapolate and generalise results that apply to a specific ecosystem. Regulating hydrologic services, for instance, is vital in certain situations; it is less so in other locations and may even be completely irrelevant in others.

The 2009 report "Climate Issues Update" ends with five suggestions that could serve as a road map for the future:

- increase benefit feedback via PES and the market exploitation of "green" products;
- reform public sector subsidies that encourage, either directly or indirectly, the use of fossil fuels;
- improve economic regulations, in terms of pricing and compensation, by applying the polluter pays principle and by taking into account the full costs—including environmental costs;
- supporting protected areas (marine, coastal and terrestrial) on which one-sixth of humanity depends for food and survival;
- invest in ecological infrastructures when the investments are less costly than restoring degraded ecosystems, and the social benefits are several times higher than the costs.

The 2010 report "Ecological and Economic Foundations" completes the analysis and provides national and local policy makers with a set of guiding questions on ecosystem services: What are the most important ecosystems in my situation? Which human communities are most dependent upon them? Which services are threatened due to current practices and policies? What changes should be expected from such and such a policy?

The Importance of Space: Bringing Together Landscape and Territory

Just as the agronomic and ecological perception of ecosystem services has led to moving from a local analysis—the scale of the plot—to a vision that includes the landscape, the socio-economic perception reflects the concept of territory. While agronomists and ecologists define the landscape through the relationships of ecological processes, territory is essentially determined by the interconnectedness of human and non-human activities with ecological processes. The integration, or rather joining together, of these two methods of representing space is a key point in research relating to the relationships between agriculture and ecosystem services. It encourages research organisations to refocus their attention to regional structures that are common to the study. In France, for example, this is done through "workshop zones", created by the French National Centre for Scientific Research (CNRS), which bring together universities and other research organisations. The disciplines remain too segregated, with each one attacking the problem of space according to its own definition. Nevertheless, there has been significant progress made over the past few years: recent work on tropical forest management [22] or the ecological functioning of grasslands and farming systems [23] show that pluridisciplinary integration is possible, and, most notably, productive.

The Communicating Vessels Principle: Land Sharing or Land Sparing?

The connectivity of ecosystem functions and services is expressed both on the landscape and territorial levels; their organisation and management are a result of preferences and hierarchies of the players at these scales, preferences and hierarchies which are influenced by national and international benchmarks. At these upper levels, the same organisational and management problems arise. The concept of an agricultural system that is both productive and ecological—detailed in Chap. 8—refers back to the optimisation of ecosystem provisioning services, especially the provision of food products, without harming the other ecosystem services. The term ecological intensification is also used, according to the expression coined in France by Michel Griffon [24]. This type of agriculture focuses on the converging of biodiversity processes and the provision of ecosystem services on the landscape scale (as defined by agronomists and ecologists). Ecological intensification is thus a result of production methods and the way space is used. Pushed to the extreme, that is, to a planetary scale, the concept can be extended in a somewhat abusive manner to recommend using the same hierarchy of ecosystem services everywhere. This leads us to the concept of "land sharing". It should be noted, however, that current work in this area is often limited to a more local vision [25], and makes generalisations on a global scale without sound analysis of the consequences [26].

As previously shown, the hierarchy of ecosystem services varies in time and space. Extreme caution must be taken when applying local generalisations on a global scale. Thus, when applied to agricultural product volume (in tonnes) and not to surface area (in hectares), the increase in yields in the United Kingdom between 1961 and 2005 showed GHG savings of 161 Gt of equivalent CO_2, well above what could have been expected from alternative mitigation strategies [27]. However, today there are no analytic studies that take into account ecosystem services in their entirety and their interactions. The study by Burney et al. mentioned above only looks at two services: agricultural production and regulating climate change. This research assuages a more differentiated vision of the hierarchy of services, one that can be summed up by the expression "land sparing" [28]; it is no longer about practicing similar sustainable farming strategies in all areas, but rather differentiating priorities according to the ecosystem, landscape and territory. This vision has garnered attention, not only from researchers [29], but also politicians. For instance, the 2010 report by the European Environment Agency, "Assessing Biodiversity in Europe" [30], underlines the importance of establishing an "optimal" relationship between protected and productive zones, whether they are used for food, energy or other goods. There are at least four challenges that must be addressed simultaneously on a worldwide scale: feed an expanding population using less arable land per capita, provide that population with energy, preserve the environment, and reduce development inequalities [9]. While it may be risky to prioritise them, it is clear that these challenges depend on the ecological conditions and levels of social and economic development within a population. It is therefore necessary to establish a conceptual framework that enables environmental consequences of activities by poor populations to be distinguished from those of rich populations, as well as distinguishing the services from which they benefit [31].

The optimum use of ecosystem services is liable to play a vital role in the major decisions of territory and land use conversion, on local, regional, national and global levels. It is no longer just about preserving the huge, emblematic rainforests of the Amazon or the Congo to better handle the issues of climate change and GHG emissions. The supporting, provisioning, regulating and cultural services in all ecosystem types must also be an integral part of any solution. To reach sustainable solutions, feeding the planet must be part of a heightened understanding and improved management of these services [32]. As for biodiversity, the notion of ecosystem services demands clarification of the interfacing between the scientific vision and that of the policy makers with society's interests at stake. By definition, ecosystem services place people and their actions in the centre of the universe. This vision is often rejected by those that believe that man is just one link in the chain of living organisms. For researchers, this opposition is irrelevant, because the concept is a tool, a way of representing a part of the ecosystem reality and functioning. Integrating ecological, agronomic and socio-economic knowledge—something that very much needs to happen with regards to ecosystem services—only makes sense if it is organised according to the questions asked. Thus, a "simple" assessment of the importance of services disturbed by a profound restructuring of an ecosystem's use—a forest, a mangrove—suffices to give weight to a policy decision. This assessment, even if imperfect, is generally possible based on available scientific knowledge. This is not the case for progressive changes in land use that concern, for example, cultivated areas, wetlands or permanent grasslands, or the shift in agricultural practices and systems. The impact of these changes is subtle and requires detailed ecological analyses using methods that are still in the dawn of their development. Regardless, from a macroeconomic viewpoint, giving value to ecosystem services is essential to sustainably managing their use.

References

1. MEA—Millenium Ecosystem Assessment (2005) Global Assessment Reports; Volume 1: Current State & Trends; Volume 2: Scenarios; Volume 3: Policy responses. Island Press, Washington, D.C.
2. Gomez-Baggethun E, Mingorria S, Reyes-Garcia V, Calvet L, Montes C (2010) Traditional ecological knowledge trends in a transition to a market economy: empirical study in Dõnana Natural areas. Conserv Biol 24:721–729
3. Costanza R, d'Arge R, de Groot R, Farber S, Grasso M, Hannon B, Limburg K, Naeem S, O'Neill RV, Paruelo J, Raskin RG, Sutton P, van den Belt M (1997) The value of world's ecosystem services and natural capital. Nature 387:253–260
4. Le Roux X, Barbault R, Baudry J, Burel F, Doussan I, Garnier E, Herzog F, Lavorel S, Lifran R, Roger-Estrade J, Sarthou JP, Trommetter M (eds) (2008) Agriculture et biodiversité. Valoriser les synergies. Expertise scientifique collective, synthèse du rapport, Inra (France), 113 p

5. Amigues JP, Debaeke P, Itier B, Lemaire G, Seguin B, Tardieu F, Thomas A (eds) (2006) Sécheresse et agriculture. Réduire la vulnérabilité de l'agriculture à un risque accru de manque d'eau. Expertise scientifique collective, synthèse du rapport, Inra (France), 72 p
6. Butault JP, Dedryver CA, Gary C, Guichard L, Jacquet F, Meynard JM, Nicot P, Pitrat M, Reau R, Sauphanor B, Savini I, Volay T (2010) Ecophyto R & D. Quelles voies pour réduire l'usage des pesticides? Synthèse du rapport d'étude, Inra (France), 90 p
7. Arrouays D, Balesdent J, Germon JC, Jayet PA, Soussana JF, Stengel P (eds) (2002) Contribution à la lutte contre l'effet de serre. Stocker du carbone dans les sols agricoles de France? Expertise scientifique collective. Synthèse du rapport, Inra (France), 32 p
8. Glendining MJ, Dailey AG, Williams AG, van Evert FK, Goulding KWT, Whitmore AP (2009) Is it possible to increase the sustainability of arable and ruminant agriculture by reducing inputs? Agric Syst 99:117–125
9. Guyomard H (2009) Nourrir la planète de façon durable est possible, à condition que…. Politique étrangère 2:291–303
10. Figuières C, Guyomard H, Rotillon G (2006) Le développement durable: que peut nous apporter l'analyse économique? Académie d'agriculture de France et Inra, L'Inra: 60 ans et des perspectives renouvelées, Séance exceptionnelle de l'Académie d'agriculture de France organisée à l'Occasion du 60e anniversaire de l'Inra, 11 octobre 2006
11. Dominati E, Patterson M, Mackay A (2010) A framework for classifying and quantifying the natural capital and ecosystem services of soils. Ecolog Econ 69(9):1858–1868. (See also: Stengel P et al 2009. Le Sol. Dossier Inra. Éditions Quae, Paris)
12. Kibblewhite MG, Ritz K, Swift MJ (2008) Soil health in agricultural systems. Philos Trans R Soc Lond B Biol Sci 363(1492):685–701
13. Rands MRW et al (2010) Biodiversity conservation: challenges beyond 2010. Science 329(5997):1298–1303
14. http://www.sfecologie.org/
15. http://www.fondationbiodiversite.fr/
16. Prugh LR, Hodges KE, Sinclair RE, Brashares JS (2008) Effect of habitat area and isolation on fragmented animal populations. Proc Natl Acad Sci U S A 105:20770–20775
17. De Bello F, Lavorel S, Diaz S, Harrington R, Cornelissen HC, Bardgett RD, Berg MP, Cipriotti P, Feld CK, Hering D, Martins da Silva P, Potts SG, Sandin L, Sousa JP, Storkey J, Wardle DA, Harrison P (2010) Towards an assessment of multiple ecosystem processes and services via functional traits. Biodivers Conserv 19:2873–2893
18. Diaz S, Lavorel S, de Bello F, Quetier F, Grigulis K, Robson M (2007) Incorporating plant functional diversity effects in ecosystem service assessments. Proc Natl Acad Sci U S A 104:20684–20689
19. Bayon R, Jenkins M (2010) The business of biodiversity. Nature 466:184–185
20. Gallai N, Salles JM, Settele J, Vaissière B (2009) Economic valuation of the vulnerability of world agriculture confronted with pollinator decline. Ecolog Econ 68:810–821
21. Bielsa S, Chevassus-au-Louis B, Martin G, Pujol JL, Richard D, Salles JM (2009) Approche économique de la biodiversité et des services liés aux écosystèmes. Contribution à la décision publique. Rapport du centre d'analyse stratégique (CAS), France, 399 p
22. Baraloto C, Marcon E, Morneau F, Pavoine S, Roggy JC (2010) Integrating functional diversity into tropical forest plantation designs to study ecosystem processes. Ann For Sci 67:303–312
23. Girard N, Duru M, Hazard L, Magda D (2008) Categorising farming practices to design sustainable land-use management in mountain areas. Agron Sustain Dev 28:333–343
24. Griffon M (2006) Nourrir la planète. Odile Jacob, Paris, 456 p
25. Midmore DJ (1993) Agronomic modification of resource use and intercrop productivity. Field Crop Res 34:357–380
26. Smith P, Martino D, Cai Z, Gwary D, Janzen H, Kumar P, McCarl B, Ogle S, O'Mara F, Rice C (2007) Policy and technological constraints to implementation of greenhouse gas mitigation options in agriculture. Agric Ecosys Environ 118:6–28

27. Burney JA, Davis SJ, Lobell DB (2010) Greenhouse gas mitigation by agricultural intensification. Proc Natl Acad Sci U S A. doi:10.1073/pnas.0914216107
28. Green RE, Cornell SJ, Schalermann JPW, Balmford A (2005) Farming and the fate of wild nature. Science 307:550–555
29. Mattison H, Norris K (2005) Bridging the gaps between agricultural policy, land-use and biodiversity. Trends Ecol Evol 20:610–616
30. European Environment Agency (2010) Assessing biodiversity in Europe. European Environment Agency Report N° 5/2010, 64 p
31. Wunder S (2008) Payments for environmental services and the poor: concepts and preliminary evidence. Environ Devel Econ 13:1–19
32. Kok MTJ, Tyler SR, Prins AG, Pinter L, Baumüller H, Bernstein J, Tsioumani E, Venema HD, Grosshans R (2010) Prospects for mainstreaming ecosystem goods and services in international policies. International Institute for Sustainable Development and Netherlands Environmental Assessment Agency, 92 p

Chapter 7
Will There be Enough Land?

Land supply is an age-old concern, with echoes in the Malthus theory. With the two founding premises of his theory, which he viewed as laws of nature, namely that "food is necessary to the existence of man" and "the passion between the sexes is necessary", Robert Thomas Malthus observed that "population, when unchecked, increases in a geometrical ratio. Subsistence increases only in an arithmetical ratio", which signifies that "the power of population is (…) superior to the power in the earth to produce subsistence for man" [1]. Economist Daniel Cohen notes that "any improvement in a population's standard of living sets off an exponential demographic growth that sooner or later must be stopped for lack of available land" [2]. Malthus' theory was—and still is—highly controversial, not only in terms of his founding ideas but for its moral and political implications. Nevertheless, Malthus's "ominous" economic science, developed at the beginning of the nineteenth century when the global population was only *1* billion, had a profound influence on classic economists and led to many pessimistic ideas about the ability of the Earth to feed an expanding population. A century and a half later, the Club of Rome[1] commissioned the 1972 Meadows Report, written by a team of scientists from the Massachusetts Institute of Technology (MIT), a highly reputable university in the world of academics. This report placed new emphasis on the constraints generated by the limited character of natural resources. It outlined the dangers—especially environmental—of the world's demographic and economic growth where food and energy resources are limited.

The issue raised in the Meadows Report—how to meet the food and energy needs of a world in full demographic and economic expansion, while also respecting the environment—is more relevant today than ever before. Soaring agricultural prices in 2007 and early 2008, and again in the second half of 2010 and early 2011, brought the problem of global food security to the fore. But more specifically, it made clear

[1] The Club of Rome is a thinktank that brings together scientists, economists and officials from national and international institutions, as well as industrial players to deal with complex issues that all countries, whether developing or industrialised, must face. It takes its name from its first meeting, held at the Accademia dei Lincei in Rome on 8 April 1968. The Meadows Report was named after two of its co-authors, Donella Meadows and Dennis Meadows (the other two co-authors were Jorgen Randers and William Behrens).

M. Guillou, G. Matheron, *The World's Challenge*,
DOI 10.1007/978-94-017-8569-3_7, © Éditions Quæ, 2014

the issue of land scarcity in a world where agriculture is expected to not only produce food, but also industrial goods—especially bioenergy—to be substituted for petrochemical products (see Chap. 5), as well as ecological services that are difficult to put a price on (see Chap. 6). This results in questioning both biofuels and meat-based products, which both require excessively large amounts of space to produce. The potential impact of climate change on land available for crops also contributes to the issue of land scarcity. Land grabs in the poorest countries in the world—which happen to be rich in cultivatable or potentially cultivatable land—have increased awareness of this issue in the media, society, and economic and scientific circles.

Agricultural Land: A Look at the Current Situation

Of the 13 billion ha (ha) of the world's land area, nearly 38 % (4.9 billion ha) is used for agriculture and 30 % (3.9 billion ha) is covered by forest. The remaining hectares constitute land unsuitable for agricultural production or forest uses, or which is difficult to convert to such uses for various climatic, topographic or pedological reasons; inland water (lakes, ponds, rivers and other underwater lands); and artificial surfaces (cities, roads, etc.). Nearly one-third of agricultural land is allocated to annual or permanent crops (1.6 billion ha) and the remaining two-thirds to permanent meadows and pastures (3.2 billion ha). Total land area equipped with irrigation systems amounts to 287 million ha, a relatively modest amount that has nevertheless doubled over the past 50 years, and which has been an essential driving force in agricultural production growth over that period[2].

Slight Increase in Agricultural Land over the Past Fifty Years

Between 1961 and 2007, worldwide agricultural land area increased by 476 million ha, mostly to the detriment of forest areas that were reduced by around 400 million ha. The expansion of agricultural land is divided between crops (just over one-third, 183 million ha) and meadows and pastures (slightly less than two-thirds, 293 million ha). Consequently, the percentage of land allocated to crops grew slightly more (12 %) than land for meadows and pastures (9 %). Two other points should be noted: first, the marked contrast between the amount of land used for crops between developed countries (continual decline over 50 years) and developing countries (continual increase over 50 years) [3]; and second, the slight decline (33 million ha) in worldwide agricultural land area since 2001, due entirely to the contraction of permanent meadows and pastures as land used for crops continued to increase, albeit at a slower pace than over the previous four decades from 1960–2000 (Table 7.1).

[2] The data in this section, unless otherwise indicated, are from the FAOSTAT database, dating from 2007 (online database consulted on 10/08/2010).

Table 7.1 World land use (Source FAOSTAT database)

	Billions of hectares	%
Total land area	13.1	100
Agricultural land, including		
Annual and permanent crops	4.9 1.6	38
Permanent meadows and pastures	3.2 12.3	25.7
Forest area	3.9	30
Land which is unsuitable for agricultural production or forest uses or which is difficult to convert to such uses, inland water, artificial surfaces	4.1	32

Is There Still Potentially Cultivatable Land that has not yet Been Cultivated?

The International Institute for Applied Systems Analysis (IIASA) has created a system to evaluate potentially cultivatable rain-fed (non-irrigated) land, based on five suitability categories: very suitable, suitable, moderately suitable, marginally suitable and not suitable [4]. The potential of currently cultivated land can then be assessed to determine how much land area could be expanded for additional cultivation. According to assessments by Fischer et al. [4], potential arable rain-fed land stands at 4.2 billion ha, but its distribution is uneven throughout the world. This land is very abundant in Latin America (1066 million ha, or 25 %), sub-Saharan Africa (1031 million ha, 25 %) and in industrialised countries (874 million ha, 21 %). There is much less in East Asia (366 million ha, 9 %), South Asia (220 million ha, 5 %) or North Africa and Central Asia (99 million ha, 2 %). Of these 4.2 billion ha that are suitable to rain-fed cultivation, 1.6 billion ha are already in crop production. That leaves 2.6 billion potentially cultivatable ha, most of which—1.8 billion ha—are located in developing countries. Almost two-thirds of suitable rain-fed land is found in 13 countries, all in Latin America or sub-Saharan Africa, with the exception of Indonesia. These countries are, in descending order: Brazil (598 million ha), the Democratic Republic of Congo (188), Argentina (104), Sudan (94), Indonesia (90), Angola (81), Colombia (72), Bolivia (67), Tanzania (65), Mozambique (62), Venezuela (61), Peru (58) and Zambia (57) [5]. This favourable outlook seems to indicate that cultivatable land may not be a limiting factor. Furthermore, most of this potential arable land is classified as very suitable (32 %) or suitable (36 %) to rain-fed agriculture. This optimistic outlook does, nevertheless, merit some discussion.

Taking Current Uses of Potential Arable Land for Rain-fed Agriculture into Account

For Fischer et al. [4], all land which has the minimum required agronomic properties (rainfall, temperature, soil, etc.) is considered potentially arable. These potential arable surfaces are therefore either already cultivated or used for other purposes (grasslands, forests, urban uses, etc.). Of the 2.6 billion ha of land that are suitable for rain-fed agriculture but not yet used for agricultural production, Nachtergaele and George [6] estimate that cities and urban infrastructure cover 60 million ha, protected zones 200 million ha, and forests 800 million ha. This means that the potential gain is limited to permanent grasslands and pastures, or 1.5 billion ha. Alexandratos [7] takes a more pessimistic view: based on similar calculations, only 1 billion ha can be taken from permanent meadows and pastures, two-thirds of which are located in developing countries.

In three scenarios with increasingly restrictive conditions on suitable land, Roudart and Even [8] also take into account land quality. In the first scenario, all land classified to some degree as suitable for agriculture (very suitable, suitable, moderately suitable or marginally suitable) can be used for crops, regardless of its current use. The amount of cultivated land can therefore be expanded by 2.35 billion ha. In the second scenario, land meeting these criteria can be used for crops, if not currently covered by forests; the amount of cultivated land can therefore be expanded by 1.45 billion ha. Finally, in the third, most restrictive scenario, only land from the first three classifications can be used for crops, unless covered by forests. In this case, cultivated land can only be expanded by 1 billion ha.

Fischer and Shah [9] follow a similar line, restricting potential arable land not yet in crop production to only non-wooded areas, in zones where population density is below 25 inhabitants/km^2. This leaves a cultivatable land reserve of 445 million ha, half of which is located in only six countries, all in Africa (Sudan, the Democratic Republic of Congo, Mozambique, Madagascar, Chad and Zambia). To sum up, when urban zones, protected areas and forests, as well as land quality, are all taken into account, the amount of potential arable land decreases. There is still a great amount of potential land, but taking land from permanent meadows and pastures to grow crops demands a simultaneous analysis of the needs for animal products and changes to be made in livestock farming. This issue was touched upon in studies focusing solely on determining potential arable land.

Uncertainty in Methodology

The methods used to assess the maximum potential of cultivatable land lead to uncertainty about the results. The first uncertainty is a result of cross referencing different, imperfect databases on land use. Determining current land uses is highly dependent upon the resolution quality of the satellite images used, the label given when plant cover was observed (forest, savanna, plantation, shade cropping, agroforestry, etc.), and the inclusion of various vegetation covers in the "potentially arable" category, etc. Moreover, FAO data on agricultural land may also be inaccurate and are very dependent upon the reliability of information provided by each country. Certain hypotheses and calculation rules can lead to underestimated potential. The analysis is static and therefore does not take into consideration any land improvements or developments that would allow land categorised as not suitable to be reclassified as suitable to some degree for agricultural production [8].

In certain Asian, North African and Central Asian countries, human intervention, by way of terracing slopes or irrigating dry land, means that cultivated land today exceeds the potential arable land total [10–12].

Conversely, other theories lead to overestimating potential arable land. Land is classified into a given category when one of the plant species being considered can reach the yield range that defines that classification. This can lead to overestimating potentially cultivatable land. For example, large areas in North Africa are considered as potentially able to grow olives or other minor crops, even though current knowledge and techniques would not make this a reality [13]. Furthermore, potential arable land often faces issues of low fertility, ecological fragility, toxicity (salinity), unsuitable infrastructure, etc. which limit its intrinsic productivity. It would require major amounts of additional water, fertiliser, and plant protection products, and would need appreciable human skills to exploit them. In some cases, costly investments would be in order to ensure accessibility and safety. According to Fischer et al. [4], more than 70 % of potential arable land in Latin America and sub-Saharan Africa—where land reserves are the greatest (see above)—would be affected by at least one of the constraints listed above, not to mention problems related to property rights and land use in numerous countries.

One should keep in mind that the analysis up to this point is limited to technical aspects; in other words, it only looks at land that could hypothetically be cultivated from an agronomic standpoint. Any conclusion must naturally integrate the economic, social and political dimensions that determine land use and its conversion, as well as who stands to win—or lose—from these changes. Additionally, losses in land that is currently cultivated must also be considered, both from a quantitative (reclassification to not suitable because of erosion, high salinity, etc.) and qualitative standpoint (lowered classification). It is also worth recalling that this analysis deals only with the suitability of land to rain-fed agricultural production, omitting the possibility (or impossibility) of extending land area via irrigation. Finally, the impact of climate change should certainly be analysed. These various points will be dealt with later in this chapter. Despite the uncertainty mentioned above, it seems that arable land reserves that are not yet in agricultural production do exist, without the need to encroach on forest land, protected zones and land used directly by people (cities, transport infrastructures, etc.). If any hard and fast figure should be remembered, it should be the 1 billion ha—reserving a wide margin of error—of possible land expansion to the detriment of permanent meadows and pasture. Converting suitable land that is currently under forest cover would bring this figure to 1.5 billion ha, a large part of which are located in South America and sub-Saharan Africa.

The Impact of Climate Change on Agricultural Production Potential

The impact of climate change on agricultural production can be broken down into effects on yield and effects on land area. Here, the discussion will focus mainly on yield, but will touch on how the effects on land area may or may not compensate

Table 7.2 Climate scenarios analysed by Fischer et al. 2002 [4] to assess the impact of climate change on agricultural potential

	Temperature increase	Precipitation change
Uniform increase in each month of both minimum and maximum temperatures		
Scenario 1	+1 °C	X
Scenario 2	+2 °C	X
Scenario 3	+3 °C	X
Uniformly distributed combined increases of temperature and precipitation		
Scenario 4	+1 °C	+5 %
Scenario 5	+2 °C	+5 %
Scenario 6	+2 °C	+10 %
Scenario 7	+3 °C	+10 %

for anticipated changes. The effects on land area will be more thoroughly discussed in Chap. 8.

Impact on Potential Arable Land

The work carried out by Fischer and his colleagues on this subject is an internationally recognised reference. In 2002, they studied the impact of seven climate change scenarios on rain-fed wheat crops. These scenarios included uniform increases in both minimum and maximum average temperatures, combined or not with precipitation increases. The results are shown in Table 7.2 above.

The simulated climate changes shown here would have an overall positive impact on land that is potentially suitable for wheat cultivation, with yield increases ranging from +1.4 % (Scenario 3) to +8.9 % (Scenario 6). In each scenario, the worldwide average conceals significant regional differences, but reveals nonetheless two consistencies:

- an increase in arable land in developed countries, from +10.7 % in Scenario 1 to +23 % in Scenario 7, despite decreases in Western Europe and Oceania;
- a decrease in arable land in developing countries, from −12.5 % in Scenario 1 to −34.4 % in Scenario 3, particularly in South America and sub-Saharan Africa.

Because it does not take into consideration crop substitutions, this analysis overestimates both increases and decreases in yield. For major rain-fed cereal crops (wheat, rice and maize), the seven scenarios applied to land classified as very suitable, suitable and moderately suitable show usable land gains ranging from +1.1 % (Scenario 3) to +5.9 % (Scenario 6). The possible expansion of land in developed countries (from +11.1 % in Scenario 1 to +25 % in Scenario 7) appears to be consistently greater than constriction in developing countries (from −1.3 to −11.1 % in the same scenarios). Research by the Center for Sustainability and the Global Environment (SAGE) at the University of Wisconsin confirms these overall results. Expanding cultivatable areas would take place in cold zones in the Northern and Southern hemispheres and at high altitudes (the Andes and the Himalayas). Constrictions would happen in Africa, the northern part of Latin America, Central America and Australia. Ultimately, climate change should have a slightly positive impact on land suitable to major cereals by 2050. But the most significant effect would be an altered geographical distribution of that land. It would worsen the situation of zones that are already at a disadvantage today, and favour exporting zones; it would also increase the need for exchange between the exporting zones towards the disadvantaged zones.

Impact on Yields

The impact of climate change on yields is the result of different natural phenomena. It is due in part to changes in average minimum and maximum temperatures and rainfall regimes and variability. It is also a result of increasing CO_2 in the atmosphere, which is conducive to photosynthesis and promotes better water efficiency in plants. Finally, it is also a question of the ability of farmers to both adapt the species and varieties they grow and change their farming practices. Fischer [14] examined four scenarios, for the 2050 and 2080 timelines, each more optimistic than the previous in terms of the fertilising action of CO_2 and farmers' ability to adapt[3].

By 2050, climate change could only have a highly positive outcome if the fertilising effect of CO_2 was offset by an effective adjustment on the part of farmers, notably by changing species grown. Short of this, productivity would diminish by 5 % in developing countries and 3 % in developed countries. The estimations are more pessimistic for 2080, when, according to the median forecasts by the UN, the world population will reach an all-time high. The gain reflected in the most optimistic scenarios would be cancelled out. The deterioration compared to 2050 would affect all countries; losses in North Africa, for example, would go from −8 % in 2050 to −23 % in 2080.

When the impacts on cultivated land and yield are combined, the result is an uncertain, although rather modest in terms of average and relative values, outlook of the potential consequences of climate change on yield for 2050. In the most optimistic hypothesis, yields could increase by around 6 %. Any other scenario either reduces or cancels out this gain. After a longer period of time, the situation would worsen. Certainly the most salient point of these scenarios is the impact for certain regions that would see their potential in cereal production affected by lower yields. These lower yields would not be offset, and would perhaps even be aggravated, by the change in exploitable land. This is especially true for sub-Saharan Africa.

[3] An effect according to which, due to increased CO_2 in the atmosphere, a greater quantity of CO_2 is absorbed and transformed into biomass. This same effect was observed between 1982 and 1999. Nevertheless, according to Zhao and Running (2010), over the 2000–2009 period, global warming most likely had the inverse effect on CO_2 storage by plants, as the negative impact of water stress weighed more heavily. The two scientists remain cautious, refusing to come to any hasty conclusions about whether it was just an anomaly or a reversed trend. For more details, see Zhao M., Running S. 2010. Drought-Induced Reduction in Global Terrestrial Net Primary Production from 2000 through 2009. Science 329(5994): 940–943.

Non-food Uses for Biomass: An Excessive Need for Land?

In 2007, worldwide CO_2 emissions linked to petroleum-derived products and their uses amounted to 10.9 gigatonnes (Gt)[4]. The transport sector produced 60 % of this amount, or 6.6 Gt, of which the majority (4.8 Gt) was from road transport. On a short- or medium-term basis, excluding given energy savings and automobile technology, biomass-derived biofuels appear to be the main, if not only, alternative to oil in transport, especially for road transport. Energy uses of renewable resources should rise sharply over the next decade, and even further by 2050, due to higher energy demand and the progressive scarcity of non-renewable energy resources. Food demand will also be higher. This leads to the question of whether the Earth can simultaneously satisfy food and non-food (mostly energy) needs, in a context where development must be sustainable, or at the very least, significantly more sustainable than it is today. In this chapter, the question will be examined from the viewpoint of competition between food and non-food—mainly energy—land uses. Chap. 5 deals specifically with the other problematic aspects of non-food biomass uses, particularly with regards to outlooks and research needs.

Criticism of First-Generation Biofuels

The strong growth over the past few years in non-food uses—namely, biofuels—for agricultural products has been the subject of scathing criticism. The environmental, energy and economic assessment of biofuels is not as positive as originally hoped, and the development of these sectors poses a threat to world food security. In both cases, one key variable is land use conversion. The first-generation biofuels on the market today are produced from raw material derived from terrestrial plant reserve organs (seeds, roots) that are also used for human and animal food (see Chap. 5). The fear of excessive competition between food and non-food uses is a reality. In recent years, this fear reached its apogee when agricultural prices skyrocketed at the end of 2007 and early 2008, and lingers on today given market prices at the beginning of 2011. It will not ease in the future, when more than 9 billion people in 2050 will need to be fed, all the while respecting the environment and natural resources.

This Malthusian fear is compounded by concerns that the environmental assessment of first-generation biofuels and their ability to reduce greenhouse gas emissions could be negative when land use conversions are factored in. The GHG assessment of a first-generation biofuel is actually positive, although it can vary widely depending on the raw plant material used as input when a given land area is analysed. It is much less so, and even negative, if producing a hectare of a sugar, cereal or protein-oil crop implies taking over a hectare of grassland or clearing a hectare of forest [15, 16]. In practice, the estimation of a given land area is based only on the environmental benefit, i.e., the reduction of GHG emissions by replacing a litre of

[4] Worldwide CO_2 emissions for all sectors combined amounted to 29 Gt in 2007 (1 Gt equals 1 billion t).

fossil fuel with the same volume of biofuel. This calculation should naturally be extended by taking into consideration the environmental costs. These costs are mainly due to land use conversion, or more precisely the loss in carbon storage provided by grasslands and forests, when 1 ha of land used for an energy crop requires, either directly or indirectly, sacrificing a hectare of grassland or forest.[5] Moreover, the instantaneous effect linked to meadow conversion or forest clearing is compounded by a significant loss related to unrealised biological production and carbon storage.

The environmental aspect cannot be reduced to simple GHG emissions. Losses in biodiversity, the use of fertilisers and pesticides beyond the absorption capacity of the environment, etc. also have potential negative effects stemming from converting land for energy crop cultivation that should also be considered. Additionally, the energy efficiency of first-generation biofuels per land unit used is quite modest, with the exception of bioethanol derived from sugar plants, generally sugarcane. Lastly, production costs may be high, especially when the prices of the raw plant materials used are high. At a time when questions—or criticisms—are being raised, an increasing number of people are asking to put an end to first-generation biofuels production, or at least an assurance that their development will not be to the detriment of food calories and upon the condition that environmental, energy and economic assessments are positive. Moving beyond the scientific and technical progress that could be made on these three points, long-term hopes (discussed in Chap. 5) deal with newer generations of biofuels.

Hopes for the Next Generation of Biofuels

Although first-generation biofuels are obtained from the reserve organs of certain crops, second-generation biofuels are produced using lignocellulose. As a reminder, the three principle sources of lignocelluloses are (a) residues and waste from agricultural, forestry, industrial, urban and household sources; (b) wood from forests; and (c) dedicated annual or perennial plants. When the raw material used to produce a second-generation biofuel is a residue or waste product, competition with food use is not a direct concern. However, it should be noted that these residues and waste products are, at least in part, used today for animal feed or to cook food. The potential negative impact on microbiological and physical soil properties due to removing agricultural and forest residues should also be mentioned.

Powlson et al. [17] conclude that energy savings linked to ploughing wheat stalks into arable land is higher than that generated by removing them for biofuel or electricity production. In practice, there are two issues brought up regarding using residues and waste for energy: first, the available biomass potential, and second, the

[5] The effect is direct if the hectare of an energy crop replaces a hectare of meadow or forest. It is indirect if the hectare of an energy crop replaces a hectare of the same crop, but meeting food needs or the economic situation, notably agricultural prices, leads to moving this hectare of a food crop to land previously used for grassland or forests (see, for example, Morton et al., who analyse the impact of high soybean prices on deforestation in Brazil: Morton D.C., Defries R.S., Shimabukuro Y.E., Anderson L.O., Arei E., del Bon Espirito-Santo F., Freitas R., Morisette J. 2006. Cropland Expansion Changes Deforestation Dynamics in the Southern Brazilian Amazon. PNAS, 1003(39):14637–14641).

cost of its use, notably the cost to collect it (because the tonnes per hectare are low and supply basins will be large) and store it.

When the raw material is a forest resource or dedicated crop—whether or not the crop could be used for food purposes—the issue of competition for land use theoretically remains the same regardless if used for first- or second-generation biofuels. Supporters of second-generation biofuels derived from dedicated crops suggest using marginally suitable land, which implies that it is unfit in some way—namely for economic reasons—for food crops. Besides the fact that the potential of marginally suitable land is not well known, at least part of these crops must be planted on good land to obtain sufficient yield to achieve minimum economic profitability[6]. To sum up, land allocation to the different uses, whether for food or non-food purposes or for recreational, environmental, or urban purposes, is also an issue for second-generation biofuels.

Biomass Energy Uses: Possibilities and Consequences

Naturally, there will be competition between food and non-food uses of biomass if there is high demand for non-food uses—and this will most likely be the case.

Sharply Rising Demand Ahead

The International Energy Agency (IEA) quantified the total energy demand for 2050 based on efforts made to reduce GHG emissions [18]. Demand would fall within a range of 15.3 billion t of oil equivalent (Btoe) for the most optimistic and restrictive scenario ("Blue Map") to 22.7 Btoe for the reference scenario if the world continues with a "business as usual" attitude ("Baseline 2050"). The Blue Map scenario limits CO_2 in the atmosphere to 450 parts per million (ppm) in 2050, equal to half the 2005 GHG emissions (and a reduction of 77% compared to the reference scenario for 2050). In this scenario, total energy biomass demand (biofuels and other energy uses) would equal 3.5 Btoe (23% of the total primary energy demand), triple the 2006 figure (1.2 Btoe). In the other two scenarios—the median "Act Map" scenario and the reference "Baseline 2050" scenario—total demand for energy biomass would also see a steep increase compared to 2006 (+1.6 Btoe and +0.8 Btoe, respectively).

High Potential for Usable Biomass

A review of literature published on the subject [19] suggests that the potential energy biomass supply would theoretically be more than sufficient to meet the world's needs. Certain research even predicts that potential offer could be *greater* than total energy demand. Smeets et al. [20] estimate the potential usable energy biomass supply at more than 35 Btoe by 2050, of which 28 Btoe could come from dedicated crops. Hoogwijk

[6] Unless costly public policies are implemented to offset low yields on marginal lands.

et al. [21] assess the potential for 2050 at just over 26 Btoe, most of which would again be derived from dedicated crops. Two other studies, one by Fischer and Schrattenholzer [22] and one by Berndes et al. [23], are much more optimistic, with a potential of over 9.6 Btoe by 2050, a figure that is nevertheless significant with respect to current energy biomass uses and the total energy demand for 2050 in the various IEA scenarios.

The high variability between the studies' estimates of potential is mainly explained by uncertainty about two key parameters: first, exploitable land for energy purposes, and second, energy crop yields (both biomass and energy yields)[7]. Exploitable land for energy purposes depends on total available land area, which must be balanced with non-energy land needs, the first of which is food, but also forest, environmental, recreational, urban and other purposes. Finally, the need for land to produce food varies according to supply (total food consumption and how consumption is split between food types) and demand (yields of the various crops). There are many unknown, or at least uncertain, factors in the equation. Just as there is high variability of overall energy biomass potential between the studies, there is also high variability of potential based on the three biomass sources, i.e. dedicated crops, forest resources, and residues and waste. Nevertheless, their relative orders of importance are generally identical. At the top of the list are dedicated crops, which equal up to 97 % of total potential for Hoogwijk et al. and at least 44 % for Fischer and Schrattenholzer; next are forest wood and residues from the forestry sector; and lastly, agricultural residues[8].

What does this mean? Essentially, despite the high variability between the assessments, these figures show that potentially exploitable energy biomass would enable the world to meet bioenergy demand in general, and demand for biofuels in particular, by 2050 and beyond. However, biomass potential was calculated on a global scale, which conceals issues of geographic distribution and the adaptation of local supply to regional demand (by country or groups of countries). Exploitable resources in abundant quantities will not always be located in high consumption areas, which means international trade will be required[9]. The potential energy biomass supply should be distinguished from the possibilities of using that biomass; in other words, the quantities of the diverse biomass used for energy must be determined along with corresponding supply and demand. The analysis cannot be limited to the bioenergy market and must be a balanced, multi-product approach, taking into consideration both alternative land uses and corresponding demand. The analysis below focuses on this issue, especially on outlining the impact of bioenergy development on global food security and the environment, measured by GHG emissions and changes in biodiversity, for the 2020 and 2050 timelines.

---

[7] Biomass yield equals the amount of dry matter produced per hectare cultivated. Energy yield is defined as the ratio of the energy value produced by one unit of dry matter (numerator) and energy value required to produce that unit of dry matter (denominator). Its calculation depends on knowing (or making hypotheses for the future) the efficiency of the processes used to transform dry matter into energy.

[8] The potential contribution of urban, industrial and household residues is rarely analysed in these global studies.

[9] By 2050, land resources, and therefore biomass production potential, will be mainly concentrated in three areas: sub-Saharan Africa, Latin America and the former USSR. In 2050, transport energy demand will come mostly from North America, Europe, East and South Asia and Russia.

One must not forget the weakness of studies that, frequently and with few exceptions (see below), seek to assess energy biomass potential without considering the consequences of making use of this potential, or which are based on unrealistic hypotheses or conjecture. Thus, the study by Fischer and Schrattenholzer published in 2001[24] is an example of the main—and sometimes sole—focus on the question of potential biomass uses for energy, with food and environmental concerns taking a back seat. This focus persists even today in much research. Furthermore, numerous hypotheses, e.g. regarding plant yield increases and the efficiency with which they can be processed into bioenergy, total increases in agricultural areas, or using crop residue for energy, have hardly been supported.

Smeets et al. analyse the outlooks for energy biomass use in 2050 based on a situation where food needs (at least, according to the authors) would first be met and deforestation would remain at base year levels [25, 26]. Food needs estimated for 2050 according to different scenarios relating to demographics and changes in food regimes are shown in terms of required plant and animal production, then with respect to necessary land area using yield hypotheses. These hypotheses depend on the technological performances of agriculture and livestock farming. The comparison with land used in 1998 results in a balance—either positive or negative—of land available for bioenergy production that can be assessed. By taking into account energy yields of dedicated crops, the supply of wood for bioenergy production, and the use of waste and residue for energy purposes, they estimate the potential of bioenergy production. Depending on the scenario, these estimates range from 8.5 to 37 Btoe, with dedicated crops being the primary source, from 59 to 82 %.

The estimates of this bioenergy production potential from dedicated crops depend on two key variables: crop yields and land availability for their cultivation. This availability increases with production yields for food, both plant- and animal-based. The most optimistic scenario permits 72 % of agricultural land, cultivated or pasture, to be freed up by 2050 compared to 1998 for energy production. Based on a hypothesis in which yields in 2050 are 4.6 times greater than in 1998 and depend on massive irrigation[10], the feasibility is questionable, to say the least. The most pessimistic scenario supposes that worldwide average yield is 2.9 times greater, and 5.6 greater in sub-Saharan Africa[11, 12]. In fact, this could actually be considered a rather optimistic outcome! The environmental consequences of changes in land allocation and production techniques are ignored, but forest cover remains constant at 1998 levels.

[10] On this point, see Berndes's analysis of the impacts on water resources on large-scale bioenergy production through 2,100. It shows that such development would aggravate water scarcity and its unequal allocation and distribution in the world. Berndes G. 2002. Bioenergy and Water. The Implications of Large-Scale Bioenergy Production for Water Use and Supply. Global Environmental Change 12(4): 253–271.

[11] In the least favourable scenario in terms of available land for bioenergy production, a part of the 1.19 billion ha of converted pasture would be used for food crops for people or animals. Consequently, the surplus of agricultural land (pasture and crop land) used for bioenergy production would "only" be 730 million ha.

[12] In all the scenarios, the supply of agricultural products enables worldwide food demand to be met. Even more surprising is that in each scenario, sub-Saharan Africa would be able to meets its food needs in 2050 using only local production by freeing up a (large) surplus of land that would then be exploitable for bioenergy production.

Dronne et al. and Dronne et al. [27, 28] evaluate two hypotheses for ethanol and biodiesel production for the 2020 timeline: production either maintains its 2006 level, or increases sharply (ethanol from 31.4 million t (Mt) to 125.1 Mt biodiesel from 5.9 Mt to 49.8 Mt. They introduce two different scenarios to replace first-generation biofuels with second-generation biofuels starting in 2015, with various hypotheses for energy yield of lignocellulosic crops. In the most optimistic case for the use and performance of second-generation biofuels (35 % of the production and 25 tonnes of dry matter per hectare), land dedicated to major "non-energy" crops in 2020 would decline from 938 to 918 million ha in order to meet increased biofuel production needs. Without second-generation biofuels, land area for such crops would only amount to 893 million ha. Developing second-generation biofuels would moderate the negative impact of biofuels on global food security, both in terms of land dedicated to major food crops and prices, but would not cancel it out. Furthermore, the decrease in by-products from first-generation biofuels would penalise the animal sector. This would not be offset by the benefits from developing second-generation biofuels, which would leave more cereals and oilseeds for food.

Beyond food security, recent research by Melillo et al. [29] and Fischer [14] integrate the environmental aspect in their GHG emissions and biodiversity calculations. Both are set at the 2050 timeline. Gains in GHG emissions from second-generation biofuels and losses due to land use changes would not balance out until this date. Melillo et al. look at two scenarios for second-generation biofuels development, both based on the objective of limiting GHG emissions. In the first, the "deforestation" scenario, all land can be used to produce food goods and biofuels provided that production is economically profitable. In the second, labelled "intensive", land that is unmanaged today, such as tropical forests, can be used only based on past land use conversion rates. The results are shown in Table 7.3 below.

In both scenarios, a large amount of forest land is converted to crop land, both for food and energy purposes, as well as pasture land in the intensive scenario. This would significantly compromise biodiversity. The two scenarios differ greatly in terms of their GHG emissions. While the first net declines compensating for land use changes in the different variations of the deforestation scenario would not be felt until 2050, they would begin to decrease before 2035 in the intensive scenario[13]. Fischer confirms and adds to this analysis for the 2050 timeline. Starting with a reference scenario that does not include biofuel production, he examines the effects of climate change in several scenarios, adds others that include biofuel production, and finally creates different scenarios if the two happen simultaneously.

The reference scenario leads to a substantial reduction in the number of people suffering from hunger—from 833 million in 2000 to 458 million in 2050—by increasing cultivated areas and yields. This number remains practically unaltered by climate change through 2050. The impacts of using biomass for energy appear to

[13] The diminished carbon storage in the soil related to biofuels production and resulting land use changes is commonly called "carbon debt" (Fargione et al., 2008. Op. cit.; Searchinger et al. 2008. Op. cit.). Over time, when land remains unchanged, this carbon debt is reduced and may even be cancelled out if the GHG emission linked to biofuels production and use are below those of the fossil fuel emissions they replace.

Table 7.3 Impacts of second-generation biofuels production based on two scenarios: deforestation and intensification (Source [29])

Scenario	Biofuels produced (Btoe)	Land area used for second-generation biofuels (Bha)	Total agricultural area (Bha)	Increase in GHG emissions as a result (Pg C)	Natural forest area (Bha) in Latin America	Natural forest area (Bha) in Africa
Deforestation	3.4	+1.48	+1.73	+103	−0.52	−0.31
Intensive	3.1	+1.39	+0.79	+34	−0.16	−0.27

The deforestation scenario assumes that all potentially arable land can be used for the production of food and biofuels. The intensive scenario assumes that land that is not currently managed may be used, but only based on the rate of change in land use observed in the past.

be much less significant. For all the hypotheses including second-generation biofuel development and the use of biofuels, from 6 to 11 % of road transport needs depending on the scenario and their variants, there is a definite increase in land area used, production and cereal prices, but the growth in demand for energy purposes reduces food uses (compared to the reference scenario of the same date). More people suffer from hunger when biofuels are being produced than when they are not. This is even more pronounced when biofuel use in transport increases and the use of second-generation biofuels (as a proportion of all biofuels) decrease. With 11.3 % biofuels included, of which 26 % are derived from lignocellulose, worldwide cereal production would increase by 313 million t and crop land by 48 million ha. Some 446 million t would be used for biofuels production and 140 million more people would suffer from hunger. This last figure suffices to show the potential reality of competition for food and non-food uses of what the Earth can produce.

In summary, the specialised studies dealing with competition between food and energy uses of land result in similar findings. The foreseeable significant increase of energy biomass demand over the coming decades can be met with a resource that seems to be abundant. But meeting that need will require land, leading to competition with other uses: food crops, temporary grasslands and permanent pastures, and even forests and other areas. If that demand is satisfied while meeting the anticipated objectives of energy specialists, there would be negative consequences for global food security, GHG emissions and biodiversity. Replacing first-generation biofuels with second-generation biofuels would ease these negative effects, and even more so if it happens rapidly using a high proportion of waste, residues or forest products.

Economic, Policy and Social Factors Affecting Land Use Conversion

The issues of expanding agricultural land and potential competition between food and non-food uses of biomass cannot be taken out of their economic, policy-related and social context. In addition to analysing potential, the factors that determine land use conversion must also be studied. This aspect is examined in light of two recent events: first, the hike in agricultural prices at the end of 2007 and early 2008; and second, land acquisition in various countries.

Soaring Agricultural Prices in 2007–2008 and Biofuels

Integrating the economic aspect, here illustrated by agricultural prices, enables the effects of land use changes to be analysed. Looking back at the situation that set the stage for agricultural product price hikes in 2007 and the first few months of 2008 provides a clear example. It shows, particularly using the example of maize in the United States, how competition plays out and the weight of agricultural and commercial policies on land use changes. The importance of the economic aspect can be demonstrated by comparing the increase in worldwide agricultural land area in 2007 and 2008 (more than 15 million ha each year) to the growth of agricultural land between 1976 and 2006 (only 2.5 million ha per year over this time period). The increase in the volume demanded clearly has a positive effect on prices. Therefore, the following assumption can be made from the start: the growth of first-generation biofuels over recent years is partly responsible for the soaring commodities prices of 2007–2008. This is especially true for cereals, due to expanded ethanol production from maize in the United States, and the price of vegetable oils following an increase in biodiesel production from oilseed crops. However, things get a bit more complicated [30, 31]. Biofuels production must be considered alongside all supply and demand factors that come into play. It is also important to respect the chronology of events, as well as differentiate between products by taking into account the effects of substitution and complementary characteristics between them, with respect to both supply and demand. Otherwise, one could argue that biofuels had nothing to do with the price hike because food prices came down during the second half of 2008 and the first few months of 2009, while biofuel production remained steady.

Biofuels Take the Stage in 2003; The First Effects are Felt in Summer 2005

In 2003, worldwide bioethanol production was equally divided between Brazil and the United States. While Brazil used sugarcane for the raw plant material, maize was used in the United States. Since then, US bioethanol production has risen considerably, due to favourable economic, environmental and policy-related factors (e.g. rising oil prices, attempts to reduce GHG emissions, mandatory programmes and tax incentives to promote the use of bioethanol). The US is quickly becoming a world leader in bioethanol production (27 million t in 2008), ahead of Brazil (19 million t), and with a strong lead on third and fourth place, the European Union (3rd place with 2.2 million t) and China (4th with 1.5 million t)[14]. Given its

[14] Source: F.O. Licht (website consulted on 21/11/2010). Bioethanol in the EU is produced using wheat (46 % in 2007), sugar beets (29 %) and imports of various plant materials to be processed locally into bioethanol (25 %). Additionally, the EU (more specifically, certain Member States and especially Sweden) is a net importer of bioethanol.

leading position, a careful analysis should be carried out on how the US managed to increase its bioethanol production and the impact of that production on higher agricultural commodity prices, especially that of maize.

This was also the year that marked the beginning of significant growth in global biodiesel production. Before 2003, there was only one significant producer in the world, the European Union, which used mainly rapeseed as its raw plant material. Since then, there has been a steep rise in EU biodiesel production (threefold increase between 2004 and 2008), due to a resolute policy of ambitious incorporation objectives [32]; however, because rapeseed and sunflower cannot do the job alone, the EU must import oilseeds, vegetable oils and biodiesel. Still, it would be risky to look at the two separately, with American bioethanol and maize on one side and European biodiesel and rapeseed on the other. This is mainly because the expansion of US maize hectarage for bioethanol, particularly during the 2007/2008 crop year, came in large part to the detriment of soybean hectarage. This had an impact on soybean prices, and due to the principle of substitution and complementary characteristics, on various cereal and vegetable oil prices. Similarly, the development of EU biodiesel production derived mainly from rapeseed oil affected not only the prices of rapeseed oil, but carried over to vegetable oils in general.

The global context in 2003 was one where supply had been lagging since the early 1990s, demand had been rising for several years, and cereal and oilseed stocks declining. Oil prices had been on the rise since the early 2000s and the US dollar was falling against an increasing number of currencies. Worldwide biofuels production was up, mainly in the United States as bioethanol. Drawing from stocks and a record cereal and oilseed harvest in 2004 enabled prices to remain relatively steady in 2003, 2004 and 2005. Moreover, international prices of the two products fell slightly in 2005. At the start of 2004, oil prices began rising sharply and the dollar fell even further. The closing prices of cereal and oilseed stocks, which increased slightly in 2004, were down again in 2005 and 2006, 2 years where climate conditions were unfavourable and global cereal production declined.

The US found itself facing diminished maize stocks and sluggish maize production, due to low prices; these factors meant the US was unable to meet its own legislation requirements for bioethanol incorporation. The country looked abroad, and more specifically, to Brazil, for bioethanol and sugarcane. The international price of sugar skyrocketed in just 8 months, with the index shooting up from 128.4 in June 2005 to 189.3 in December 2005 (up 47 % in 6 months), continuing up to 254.3 in February 2006 (up 34 % in 2 months)[15]. In response to these high sugar (and ethanol) prices, land dedicated to sugarcane in Brazil increased. American maize gradually regained ground from Brazilian sugarcane. As a result, international sugar prices took a nose-dive, starting in March 2006 and continuing through 2007. It would only be at the start of 2008 that world sugar prices would rise again, as a result of climbing worldwide agricultural prices.

[15] FAO Sugar Price Index (http://www.fao.org/worldfoodsituation/FoodPriceIndex/fr/).

And so the Story Goes

Biofuels production, and more specifically, bioethanol in the United States, experienced the same situation, with only a slight shift in time, for international maize prices. Prices rose from the end of 2006 through the first quarter of 2008, as the US sought to promote national bioethanol production using domestic plant (maize) sources. In 2003/2004, the amount of maize used for bioethanol production was equal to just under 30 million t, or 11 % of total domestic maize use. In 2007/2008, that amount had increased to 76.2 million t (22.9 %), using 8 million ha out of a total of 35 million ha of maize in the US[16]. These statistics suggest a strong influence of US bioethanol production on increased prices of US maize. Because the United States is—by far—the leading global exporter of maize, the effects quickly spread internationally.

The second effect of bioethanol production in the United States was to drive up international maize prices (and, incidentally, lead to a downward adjustment of global sugar prices). It was during the 2006/2007 crop year that the increase in global maize prices was the greatest (nearly 50 %). The higher price of maize with respect to other cereals and oilseeds led producers—particularly in the US—to dedicate more land to maize, especially during the 2007/2008 crop year. Worldwide, land used to grow maize, which had declined by 3 % between 1996/1997 and 2002/2003, increased by 14 % between 2002/2003 and 2007/2008, and half of this increase took place only during 2007/2008, mostly to the detriment of oilseed production[17]. This is the third effect of bioethanol production in the US. There is no doubt that this decrease in global oilseed land area in 2007/2008 added to the pressure on oilseed prices during the 2007/2008 crop year. Between June 2007 and July 2008, the FAO global price index for oils increased by nearly 79 %[18]; by comparison, over that same period, global maize prices rose by just over 20 %, rice prices by just under 50 % and wheat prices by a little over 90 %.

This contagion effect on various cereal and vegetable oils was compounded by various factors which caused a general increase in agricultural prices (lagging

---

[16] At that time, world maize crop land was just under 160 million ha.

[17] Abbott P., Hurt C., Tyner W.E. 2008 (updated 2009). What's Driving Food Prices? Farm Foundation, Issue Report, 75, July 2008 (updated March 2009). Global land area used for soybean cultivation rose nearly 15 % between 2002/2003 and 2006/2007, mainly in South America, to meet sharply rising Chinese demand. Since the middle of the 1980s, Chinese growth in vegetable oil consumption was greater than domestic production, a differential that means China had to import nearly 10 million t of vegetable oil in 2007, of which 6 million t of palm oil and 3 million t of soybean oil. What are we to make of this observation? Too much weight must not be given to a single determinant examined alone. It must be analysed within the context that defines changes in supply and demand of various agricultural goods.

[18] According to the OECD (2008), the oil market had a significant statistical influence on soybean and palm oil prices, as well as on rapeseed oil. Soybean oil prices also affect the price of rapeseed and palm oils. For more information, see OECD 2008. Impact relatif sur les prix mondiaux des produits des changements structurels à court et long terme sur les marchés agricoles, le choc des prix des matières premières: le retournement structurel d'une tendance de prix bas? TAD/CA/APM/CFS/MD(2008)3, OECD, Paris.

supply, disappointing harvests, rising demand for food products, diminished stocks, rising oil prices, a falling US dollar, speculation on agricultural plant products, uncoordinated commercial policies meant to discourage exports and encourage imports of agricultural goods, etc.). In the specific case of vegetable oils, the concomitant development of EU biodiesel production was an additional factor—one that is significant given the tonnage involved—affecting vegetable oil prices, particularly rapeseed oil.

Several studies have attempted to quantify the role biofuels played in the 2007–2008 spike in agricultural prices. Differences in data, methodology, scenarios, simulation timelines, etc. make comparing the results somewhat complicated. Nevertheless, aside from a few (very) rare exceptions, one thing is clear: development of bioethanol in the US played a role in rising prices, first of maize, and by carry-over effect, of various cereals and vegetable oils, including, as shown in a study by the International Food and Policy Research Institute (IFPRI), rice, even though this cereal was not used to produce bioethanol. After analysing several studies and additional ad hoc research, Collins reaches the conclusion that 25–50 % of the increase in US and global maize prices between 2006/2007 and 2008/2009 crop years could be attributed to the growth in bioethanol demand in the United States [33].

Taking Economic Realities into Account

Far from being a useless detour, looking back at the situation that led to the 2007–2008 agricultural price spike shows that the issue of potential competition between food and non-food uses cannot be analysed independently of the economic context and its determinants (climate, policies, etc.) The 2007–2008 rise in global agricultural prices is the result of various converging factors, one of which is first-generation biofuels. More precisely, it is a result of short-term factors which negatively affected the volumes available on world markets (extreme weather and restrictions on exports). Certain factors also positively affected demand (falling US dollar, biofuels development, speculation and policies that encouraged imports) and production costs (rising oil prices, and input prices in general). These factors all converged within a context of over 10 years of stagnant supply and rising food demand, leading to diminishing stocks.

The abrupt turn-around that occurred in the spring of 2008 can be explained by a reversal of the global situation, with several factors that had previously played a role in higher prices now pushing prices down: favourable climate conditions in various production zones, rising US dollar, lower oil prices, and retreating speculators. Moreover, the high agricultural prices in 2006 and 2007 encouraged producers to increase land used for major crops at the expense of fallow land, forest or grasslands. They also led farmers to seek higher yields and intensify practices. In this situation, where rising supply was met with sustained food and non-food demand, the correction of agricultural prices enabled countries to end their short-term policies of encouraging imports or restricting exports. History repeated itself again starting in the summer of 2010: fires in Russia and flooding in Australia, increased oil prices, a falling US dollar, and the return of speculators. Agricultural prices shot up again. In the minimally-regulated global commodities and energy markets, it is likely that markets will experience similar episodes in the future. The issue of cultivatable land, and more generally of competition between alternative land uses, cannot be analysed only in terms of growth potential, "theoretically" useable land and the "absolute" scarcity of land. The economic aspect must figure in any analysis, as land use is determined by prices and income, which are influenced by policies and scientific and technological advances.

The Race for Land and Large-scale Agro-Industrial Projects

Potential arable land is rarely untouched and unclaimed, whether those claims are legally recognised or not. As a result, land use conversions must also be analysed, taking the losers and the winners into account, and often, the transfer of land rights [34]. Recent trends in large-scale land acquisitions demonstrate how important the economic and political aspects are in land use conversions for the owners or users of that land. Without taking an overly optimistic or pessimistic stance on the issue, this tendency should be analysed realistically, along with the associated risks and opportunities that such development entails. Rising agricultural land investment is not a new phenomenon. It does, however, have new characteristics today. Contrary to land appropriation that occurred during colonisation, today's land acquisitions are no longer achieved by force; other political and economic means are employed to take advantage of certain populations whose rights are compromised to gain the desired resources [35]. Notices of intent (NOI) have multiplied over the past 5 years. Coveted land area often reaches a million hectares per project; the country of origin and motives of the investors are quite varied. Investors are no longer limited to developed countries; today, Middle Eastern and Asian investors have also joined the race. The target countries remain mainly developing countries, particularly African countries [36, 37].

Food Security no Longer the Sole Motive

Although media attention has focused on projects aiming to improve the food security in the investor's country of origin, the motives for land acquisitions vary [35, 38]. Of course, the 2007–2008 agricultural price crisis created a certain mistrust of economic operators with regards to the ability of the international market to provide sufficient and affordable agricultural commodities. Many projects aim to ensure a supply of agricultural commodities in case of a paucity of available quantities on the international markets or high prices. They also seek to take advantage of new opportunities to turn a profit. Attracted by the new markets opening to biofuels in industrialised countries or seeking wood, numerous projects are oriented towards biomass production or the exploitation of forest resources. Certain projects reflect medium- to long-term strategies of land accumulation and speculation in the "race for land", where using cash resources to purchase "new" assets is delayed, because land has historically outperformed other financial assets. Others want to develop non-agricultural activities such as tourism, urban development or mining explorations. Finally, others are developing or seek to develop environmental activities in a broader sense, such as creating carbon sinks to benefit from carbon credits or conserving natural spaces to take advantage biodiversity credits.

From NOIs to Reality: Lower but not Insignificant Investments

The real extent of the phenomenon is difficult to quantify because of the many forms it takes and a lack of transparency surrounding the transactions. Consequently, field research on a country-wide scale needs to be carried out to understand the actual situation and state of progress on the projects. According to the NOIs, 42 million ha may have been the object of land acquisitions between 2005 and 2009, 75 % of those in Africa. Out of the 405 projects for which data on planned activities exist, 37 % plan to produce food; 21 % to grow cash crops; 21 % biofuels production; and the remainder for the development of forest plantations, hunting reserves, or livestock farming [37]. For the target countries, the sum of the NOIs reached 3 million ha in Madagascar [39], 2.7 million ha in Mozambique, 958,000 ha in Cambodia [37] and 870,000 ha in the Office du Niger area in Mali [40].

The actual extent of the phenomenon is nevertheless much less than the NOIs and media coverage would lead the public to believe. In Madagascar, the "Daewoo Project" contributed to the deposition of President Ravalomanana, and put a stop to many other projects. For example, at the end of 2009, more than a third of the 52 projects that were announced never made it past the prospection stage or were completely stopped [41]. Investment therefore lagged, but did not stop completely. At present, 150,000 ha are being sought after; this is twenty times less than initial NOIs. Similarly, in Mali, of the 870,000 ha solicited by investors, only 50,000 ha were subject to leases ratified by Malian government services. According to the World Bank, 30 % of the projects announced in or by the press have been abandoned, 20 % are only at the prospection stage and only 21 % have led to activity commencing on the land, often on an area that is smaller than originally planned. In Madagascar and Mali, less than 2 % of the land area initially sought by investors was finally used. This slowdown in foreign land investment can be explained by many converging factors: an unfavourable international economic situation (banking, financial and economic crises; plunging market prices for commodities in 2009), a lack of investor expertise in agronomy, underestimation of investment sums required for the land and its exploitation, resistance on the part of locals and governments, just to name a few.

New Investors in Agriculture

These investments are made by various economic operators, in terms of both nationality and status. Foreign investors largely dominate the field when investments are high and the land area sought is over 1,000 ha [36]. They come from South and South-East Asia, Gulf countries, Europe and Africa. The majority are private operators. New investors, few and far between until recently, include banks, pension funds and industrial players in the energy sector [42]. Another new development is that States are becoming stakeholders through public companies, sovereign funds or intergovernmental agreements. International associations for environmental protection, often associated with sponsors or company foundations, are also part of this trend, in this case with an objective of acquiring land with high ecological value. Domestic investors are also active players. For the most part, they belong to the country's economic elite and create an agricultural project as part of a diversification strategy for their activities. Some are partners with foreign investors seeking to get involved in bigger projects. Generally speaking, the land area in question is under 1,000 ha, but the increasing number of projects weighs on land use at national level [36]. In Sudan, 78 % of the 4 million ha are coveted by domestic investors [37]. In Mali, 45 % of the 870,000 ha are solicited for lease by domestic investors: the majority seek land area under 5 ha, some 15 % for land area over 1000 ha [40].

The Host Country Government's Key Role

In most cases, transactions bring investors and the government of the host country together. In many developing countries, the land rights of rural populations are not legally recognised and the land to which these rights apply fall within the State's domain[19]. The host country governments are therefore by no means passive when it comes to investments. They actively promote this type of investment in order to reap the economic benefits and compensate for reduced public development funds. They make access to the land easier, usually through lease contracts. Nevertheless, these lease contracts are often established for 99 years with the option to renew. In other words, they are practically a sale. Financial compensation offered when a deal is struck, if offered at all, is often quite low (around 1 US$ per hectare). The possibility of indirect economic benefits generally suffices to satisfy host country governments. The investor's development obligations and commitments are, however, rarely clearly specified. The return on investment per hectare varies, according to more or less realistic forecasts by the operators, from 750 to 11,000 US$ per hectare [37].

While many countries have implemented policies of decentralised land management, land access terms for investors bear witness to a shift in the other direction: a recentralising and concentration of State power. Contracts are most often negotiated in the absence of local community representatives or rural organisations. The latter are, in the best of cases, informed when the States impose economic, social and environmental impact studies [43]. In fact, in a great many projects, land is granted to investors to the detriment of the local users who are not even consulted, let alone compensated.

[19] In many African countries, the government is recognised as the only legal owner, which stems from the principle of state-owned public domains, a holdover from their colonial history.

When laws in the host country or clauses in the investment contract allow for compensation, the procedures are not always respected and only give local populations limited income in exchange for a resource that generates revenues over many years.

Varied Development Models

The organisation of production is rarely explained clearly and transparently. Most projects are for large-scale plantations, with salaried employees and mechanised processes. These models can be beneficial to the host country and local population if they are accompanied by infrastructure and services that create jobs. Depending on the planned crop, the amount of investment can vary from 450 US$ (cereals) to several thousand US dollars per hectare (sugarcane harvested mechanically). The number of jobs created can range from 10 for mechanically harvested cereals and Jatropha and up to 420 for manually harvested rubber per 1,000 ha. These figures do not suggest significant investment in infrastructure or skilled jobs created, a crucial aspect for a developing country.

These production models also carry certain risks if the land acquired for the project implies displacing local populations and related resources, such as pastures, forest resources, water, etc. Environmental risks may also emerge. Even if certain large-scale farms are able to adopt environmentally sound practices, others may adopt harmful practices or seek only to extract what they can from the land. Nevertheless, certain projects, managed by experienced agricultural companies or financial institutions outside the agricultural sector, depend, at least in part, on the participation of local farmers. The operators plan to buy agricultural materials from local producers, establishing contracts that guarantee those producers a certain purchase price and access to technical advice, inputs, and even credit. A few projects lease a part of the land to local farmers in exchange for rent or a share of the company's profits (joint venture). Such structures can provide a real opportunity for small-scale farmers, who are penalised by difficult access to credit, factors of production, agricultural production, technical information and farming advice. However, the real gain for farmers depends on their ability to obtain fair contracts, the role accorded them by the project management and the legal measures they can leverage in order to oblige investors to respect their commitments.

Stakes that go Beyond the Here and Now

The trends in agro-investment shed new light on issues that, for the most part, already exist. First, they serve as a reminder of the importance of effective and fair governance of land and natural resources, ensuring land rights of local populations, and giving those populations real negotiating power so they are able to defend those rights. Until viable land policies are established and fully implemented, and given that there are loopholes in the laws in host countries that weaken the protection of local populations' rights and the environment, at the very least information should be more transparent and local populations and institutions consulted and provided with the means to negotiate their participation in these projects and obtain a fair value for the resources made available to foreign investors. Investor promises too often lead governments, local elected officials and even some indigenous populations to sign off on the project, while investor failure to fulfil their commitments leads to conflicts that are unfavourable to all parties involved.

In broader terms, these trends raise issues of fairness and development models related to land use changes and the owners or users of the land. These models involve different types of agriculture, generally large-scale, that use family or salaried labour. However, their success cannot be allowed to come at the cost of social rights or to the detriment of the environment, nor to compromise on the objective of reducing poverty. Nevertheless, agricultural investments, both public and private, are more necessary than ever before (see Chap. 10). These land acquisitions should not be refused outright, but rather accepted on the condition that they benefit both investors and local populations. Establishing good practices is certainly necessary, but will not be enough. The right of intervention could be applied, at least in certain cases, in the name of food security. This leads back to the necessity of improved agricultural and food governance, discussed further in Chap. 10.

The availability of land that can be used for agricultural production of food goods will be a key point in meeting the food challenge over the coming decades. The environmental impacts of converting this land to grow crops, especially in terms of GHG emissions and biodiversity, must be minimised to ensure the land's sustainability. The impact of climate change on cultivatable land appears to be relatively low on a global scale; it could, however, negatively affect certain regions, namely sub-Saharan Africa and the northern part of Latin America, thereby increasing international trade needs of agricultural food products. It will not be possible to count on a significant contribution from agriculture for biofuel production without accelerated conversion of forests and grasslands unless there is strong growth in agricultural yields, a rapid development in second-generation biofuels and high production performances of those biofuels. Public policies that encourage replacement of fossil fuels, controlling GHG emissions and protecting forest or grassland ecosystems will play a major role. Market variables, which are increasingly tied to agricultural products and oil, could upset the allocation of resources, reduce food availability and cause agricultural prices to fluctuate greatly. Increased oil prices could also serve as an incentive, just as public policy did before, to encourage energy crops to be grown to the detriment of food production. Consequently, there is still no guarantee that the food security of the poorest populations will improve, especially if that security is founded solely on production growth. Controlling consumption of the richest populations (Chap. 2), reducing waste and consumption (Chap. 3) and diminishing post-harvest losses (Chap. 4) are factors that must not be disregarded. Additionally, growth in agricultural production must, as in recent years, first be based on increased yields through productive and ecological agricultural systems. This will be discussed in the following chapter.

This worrying observation should take into account the limiting effect of water availability in many regions, a point which is not sufficiently analysed in land use forecasts. Energy and environmental policies combined with market factors are not sure fire measures to meet the global food challenge, based on current trends and especially with the crisis risks they pose. Agricultural and food policies that are able to consolidate supply on the macro-regional and global scales will also be necessary. As shown by global land acquisition trends, which followed the soaring agricultural prices in 2007–2008, such policies must be anchored in and built upon sound land management. These intense reactions reveal mounting concerns over land scarcity which must be addressed in the global debate about the future of our planet. This issue will be further discussed in Chaps. 9 and 10.

References

1. Malthus TR (2005) Essai sur le principe de population. Première édition de 1798, deuxième édition de 1803. See also Alternatives économiques Poche n 021, novembre 2005
2. Cohen D (2009) La prospérité du vice, une introduction (inquiète) à l'économie. Albin Michel, Paris
3. Bruinsma J (2009) The resource outlook for 2050: by how much do land, water and crop yields need to increase by 2050? Paper presented at the FAO Expert Meeting, 'How to feed the world in 2050?', 24–26 June 2009, Rome (figure 4 p 12)
4. Fischer G, van Velhuizen HT, Shah M, Nachtergaele FO (2002) Global agro-ecological assessment for agriculture in the 21st century: methodology and results. Research Report, IIASA, Laxenburg, 154 p (For more details on the method of evaluating potentially cultivatable land)
5. Bruinsma J (2009) The resource outlook for 2050: by how much do land, water and crop yields need to increase by 2050? Paper presented at the FAO Expert Meeting, 'How to feed the world in 2050?', 24–26 June 2009, Rome (figure 3 p 11)
6. Nachtergaele F, George H (2009) How much land is available for agriculture? Unpublished manuscript cited in Bruinsma J (2009) The resource outlook for 2050: by how much do land, water and crop yields need to increase by 2050? Paper presented at the FAO Expert Meeting, 'How to feed the world in 2050?', 24–26 June 2009, Rome
7. Alexandratos N (ed) (1995) World agriculture: towards 2010, an FAO study. Wiley, Chichester, 488 p
8. Roudart L, Even MA (2010) Terres cultivables non cultivées: des disponibilités suffisantes pour la sécurité alimentaire durable de l'homme. Ministère de l'Alimentation, de l'Agriculture et de la Pêche, Centres d'études et de prospective, Analyse N° 18, mai 2010
9. Fischer G, Shah M (2010) Farmland investments and food security. Statistical annex. Report prepared under World Bank—IIASA Project
10. Bruinsma J (2009) The resource outlook for 2050: by how much do land, water and crop yields need to increase by 2050? Paper presented at the FAO expert meeting, 'How to feed the world in 2050?', 24–26 June 2009, Rome (p 11)
11. Roudart L, Even MA (2010) Terres cultivables non cultivées: des disponibilités suffisantes pour la sécurité alimentaire durable de l'homme. Ministère de l'Alimentation, de l'Agriculture et de la Pêche, Centres d'études et de prospective, Analyse N° 18, mai 2010 (p 4)
12. Bruinsma J (2009) The resource outlook for 2050: by how much do land, water and crop yields need to increase by 2050? Paper presented at the FAO Expert Meeting, 'How to feed the world in 2050?', 24–26 June 2009, Rome (p 11)
13. Bruinsma J (2009) The resource outlook for 2050: by how much do land, water and crop yields need to increase by 2050? Paper presented at the FAO Expert Meeting, 'How to feed the world in 2050?', 24–26 June 2009, Rome (p 10)

14. Fischer G (2009) World food and agriculture to 2030/2050: how do climate change and bioenergy alter the long-term outlook for food, agriculture and resource sustainability? Paper presented at the FAO Expert Meeting, 'How to feed the world in 2050?', 24–26 June 2009, Rome (final draft August 2009). All the percentages presented in this sub-section compare the results of the A2 Hadley Centre Climate Model derived from the British model HadAM3 (Pope VD, Gallani ML, Rowntree PR, Strattton RA (2000) The impact of new physical parametrizations in the Hadley Centre Climate model—HadAM3. Clim Dynam 16:123–146), based on the IPCC A2 emission scenarios (Nakicenovic N, Swart R (eds) (2000) Special report on emission scenarios. Intergovernmental Panel on Climate Change, Cambridge University Press, 570 p) with those of a reference scenario for the horizon 2050
15. Fargione J, Hill J, Tilman D, Polasky S, Hawthorne P (2008) Land clearing and the biofuel carbon Debt. Science 319:1235–1238
16 Searchinger T, Heimlich R, Houghton R, Dong F, Elobeid A, Fabiosa JF, Tokgoz S, Hayes DJ, Yu T (2008) Use of US croplands for biofuel increases greenhouse gases through emissions for land-use change. Science 319:1238–1240
17. Powlson DS, Riche AB, Coleman K, Glendining MJ, Whitmore AP (2008) Carbon sequestration in European soils through straw incorporation: limitations and alternatives. Waste Manag 28:741–746
18. International Energy Agency (IEA) (2008) Energy technology perspectives 2008: scenarios & strategies to 2050. IEA and OECD, Paris, 650 p
19. Forslund A, Gauvrit L, Guyomard H, Mora O (2011) Les utilisations non alimentaires de la biomasse et la concurrence pour l'utilisation des terres: belles promesses ou vraies craintes? In: Léridon H, de Marsily G (eds) Démographie, climat et alimentation mondiale. Institut de France, Académie des sciences, pp 152–164
20. Smeets EMW, Faaij APC, Lewandowski IM, Turkenburg WC (2006) A bottom-up assessment and review of global bio-energy potentials to 2050. Prog Energy Combust Sci 33(1):56–106
21. Hoogwijk M, Faij A, van den Boek R, Berndes G, Gielen D, Turkenburg WC (2003) Exploration of the ranges of the global potential of biomass for energy. Biomass Bioenergy 25:119–133
22. Fischer G, Schrattenholzer L (2001) Global bioenergy potential through 2050. Biomass Bioenergy 20:151–159
23. Berndes G, Hoogwijk M, van den Broek R (2003) The contribution of biomass in the future global energy supply: a review of 17 studies. Biomass Bioenergy 25:1–28
24. Fischer G, Schrattenholzer L (2001) Global bioenergy potential through 2050. Biomass Bioenergy 20:151–159
25. Smeets EMW, Faaij APC, Lewandowski IM, Tirkenburg WC (2007) We acknowledge that food production and food security must be given priority above energy crop production
26. Smeets EMW, Faaij APC, Lewandowski IM, Tirkenburg WC (2007) We acknowledge that food production and food security must be given priority above energy crop production. (cf. Footnote p 17)
27. Dronne Y, Forslund A, Gohin A, Levert F (2009) Conséquences des politiques de biocarburants de l'UE et des grands producteurs à l'horizon 2020. Institut National de la Recherche Agronomique (Inra), Unité SMART, Rapport pour le ministère en charge de l'agriculture
28. Dronne Y, Forslund A, Guyomard H (2011) Les biocarburants de deuxième génération et la compétition pour l'usage des terres. OCL 18(1):1–9
29. Melillo JM, Gurgel AC, Kicklighter DW, Reilly JM, Cronin TW, Felzer BS, Paltsev S, Schlosser CA, Sokolov AP, Wang X (2009) Unintended environmental consequences of a global biofuels program. MIT Joint Program on the Science and Policy of Global Change, Report 168, January 2009
30. Guyomard H (2008) Expliquer les évolutions des cours des matières premières agricoles: à l'impossible, nul n'est tenu! OCL 15(6):364–377
31. Forslund A, Gauvrit L, Guyomard H, Mora O (2011) Les utilisations non alimentaires de la biomasse et la concurrence pour l'utilisation des terres: belles promesses ou vraies craintes? In: Léridon Henri, de Marsily Ghislain (eds) Démographie, climat et alimentation mondiale. Institut de France, Académie des sciences, pp 152–164

32. Bureau JC, Guyomard H, Jacquet F, Treguer D (2010) European biofuel policy: how far will public support go? In: Khanna M, Scheffran J, Zilberman D (eds) Handbook of bioenergy economics and policy. Springer, pp 401–425
33. Collins K (2008) The role of biofuels and other factors in increasing farm and food prices: a review of recent developments with a focus on feed graim markets and market prospects. Report written as supporting material for a review conducted by Kraft Foods Global Inc. on 'The current situation in farm and food markets', 19 June 2008
34. Borras S, Franco J (2010) Towards a broader view of the politics of global land grab: rethinking land issues, reframing resistance. ICAS Working Papers Series N°1 (May 2010)—J Dev Stud
35. Comité technique foncier et développement (2010) Les appropriations foncières à grande échelle. Analyse du phénomène et propositions d'orientations. Comité technique foncier et développement, Paris
36. Cotula L, Vermeulen S, Leonard R, Keeley J (2009) Land grab or development opportunity? Agricultural investment and international land deals in Africa. FAO, IIED and IFAD, London
37. Banque mondiale (2010) Rising global land interest. Can it yield sustainable and equitable benefits? Banque mondiale, Washington D. C.
38. Burnod P, Anseeuw W, Bosc PM, Even MA (2010) Appropriations foncières dans les pays du Sud: bilan et perspectives. Ministère en charge de l'agriculture, Centre d'études et de prospective, Analyse n°16, Avril 2010, Paris
39. Andrianirina-Ratsialonana R, Ramarojohn L, Burnod P, Teyssier A (2011) Après Daewoo? État des lieux et perspectives des appropriations foncières à grande échelle à Madagascar. Observatoire du Foncier à Madagascar, Cirad, ILC
40. Papazian V (2010) Les investissements fonciers au Mali. État des lieux et perspectives des acquisitions foncières à grande échelle en zone office du Niger. Mémoire de master, École supérieure d'agronomie de Toulouse, Cirad, Montpellier
41. Teyssier A, Ramarojohn L, Andrianirina-Ratsialonana R (2010) Des terres pour l'agro-industrie internationale? Un dilemme pour la politique foncière malgache. EchoGéo 11, décembre 2009/février 2010. http://echogeo.revues.org/index11649.html
42. Daniel S, Mittal A (2010) (Mis)Investment in agriculture. The role of the international finance corporation in global land grabs. The Oakland Institute, Oakland
43. Vermeulen S, Cotula L (2010) Over the heads of local people: consultation, consent, and recompense in large-scale land deals for biofuels projects in Africa. J Peasant Stud 37(4):899–916

Chapter 8
The Need to Strive for Productive Yet Ecological Agriculture

Optimised Productivity of Arable Land

Sixty years of Successful Intensification

In the mid-twentieth century, the first "green revolution"[1] led to historic growth in agricultural production. It enabled the global population to multiply by 2.4, average food consumption to increase and the proportion of undernourished people in the world to decrease. These achievements were made possible through increased productivity or yields[2] of arable land, with 85 % of productivity increases attributed to increased yields and 15 % to the expansion of cultivated areas. What we call the "intensification" of agriculture enabled production to increase without a massive disruption of natural ecosystems and the often disastrous impacts this has on the environment. The grave predictions of the 1960s (such as the ones detailed in *The Hungry Future* [1] by René Dumont and Bernard Rosier) never came to pass thanks to the invention of relatively cheap industrial inputs (fertilisers, pesticides, equipment, buildings, etc.), genetic improvement of crop varieties and improvements in water management, such as irrigation and drainage. In the most industrialised countries, industrial inputs purchased for non-agricultural uses also boosted labour productivity, which increased more than ten-fold for numerous crops and livestock. The result was a large drop in the price of food for households, all while the types of foods consumed became more diverse and the percentage of household expenses dedicated to processed products increased.

[1] By extension, we use this term to include any agricultural revolution based on the use of fertilisers, pesticides, improved varieties, mechanisation and irrigation, no matter where it occurs, even in Europe.

[2] These two notions may differ if several crops are grown in the same plot, either at the same time or successively throughout the year. In some regions, increased productivity is linked to increased yields as well as the practice of multiple harvests in one year.

M. Guillou, G. Matheron, *The World's Challenge*,
DOI 10.1007/978-94-017-8569-3_8,

The Continuing Importance of High Yields

Taking into account forecasts for demographics, urbanisation and economic growth—and the consequences these factors will have on eating habits—demand for agricultural products should increase by around 70 % by 2050. Satisfying consumption needs all while substituting fossil fuels with renewable fuels from biomass in the production of industrial goods (see Chap. 5) will require maintaining an elevated rate of growth in agricultural production. This rate should be close to that of 1.5 % per year observed in the first decade of the twenty-first century. The role the expansion of agricultural land and increased yields will play is a key issue for the global environment. Global reserves of arable land seem, in theory, to be sufficient to ensure a majority of this growth; the main uncertainty involves the development of biofuels.

In Chap. 7, we estimated the amount of cultivatable land that could potentially be added at 1 billion hectares, taking into account the restrictions tied to the capacity of this land, forest conservation and protected zones. This land would mostly be "taken" from grasslands and pastures. If this figure of 1 billion hectares is correct (although reasonable caution must be taken when using this estimate), it is equivalent to approximately 70 % of the surface area of currently arable land. However, this ignores the losses that occur from the many different degradation processes of cultivated soils, arising partly from the impact of production techniques combined with natural factors. Degradation may be caused by erosion from water or wind, structural damage from working the soil, soil compaction by harvesters and fertiliser machines, reduction in the levels of organic matter, salinisation or acidification or chemical contamination. The very diversity of these processes, and their varied intensity in space and time, make it difficult to evaluate the amount of land they have rendered unusable, or the equivalent loss in production. Estimates of annual loss during the last quarter of the twentieth century vary from 6 million hectares to Robert and Cheverry's estimate of 12–16 million hectares [2]. The expansion of urban, industrial and communication infrastructure also leads to the consumption of arable land, but the extent of this impact is very hard to determine. Estimates vary between two and 19.5 million hectares lost per year, according to a special report published by the UN in 2010. According to ICIMOD's 2008 report, China lost nearly 1 million hectares per year between 1979 and 1995. It was not specified how much of this land was arable, but it is worth noting that the majority of urban settlement occurs in areas traditionally dedicated to agriculture.

A conservative view puts the total losses in arable land at 10–20 million hectares per year. Even if we can assume that this figure will remain stable in the coming decades, it significantly modifies our estimate of reserves of potentially arable land, which would automatically reduce the figure considerably. This means that soil maintenance and preservation will play a decisive role. The availability of land makes it reasonable to assume that the answer to food requirements will not reside in a large-scale expansion of cultivated areas (see Chap. 7), without affecting valuable and/or forest ecosystems. We must carefully evaluate the effects of cultivating natural ecosystems or those used more or less intensively for farming, grazing and permanent grasslands. The profound changes in the functioning of the ecosystem that could arise may alter the services it provides by affecting biodiversity and water regimes as well as incurring other risks, such as elevated carbon emissions. The ecological cost of expanding agricultural lands may also have certain positive trade-offs. A limited proportion of cultivated land in natural areas can contribute to local biodiversity, for example by creating new ecotones. Consequently, the imperative to increase the productivity of arable land on a worldwide scale will win out. The future of livestock farming, and thus of the consumption of animal products, will be a determining factor, given the percentage of plant products dedicated to animal feed. If consumption remains constant, the productivity of the land multiplied by the yield of livestock farming will determine the space needed, but this analysis must be made more specific in order to take into account the role of herbivore farming in the exploitation of areas not conducive to crop farming and in the maintenance of preserved ecosystems, such as mountain grassland.

Environmental Degradation

If we admit that agricultural production must rise, we must support increased agricultural intensification and the use of factors and methods that allow crops to grow under the most productive conditions. But the environmental costs of pursuing intensification, as was done over the last few decades, are high.

Environmental Impacts of Intensive Agriculture

The impacts—or what economists call "negative externalities"—of intensive agriculture have effects on the environment, natural resources and biodiversity.

Water and Aquatic Environments: Contamination, Conflicts of Use, Resource Depletion, Risks

We are already well aware of the consequences that arise when the quality of surface water and groundwater is affected. Excess nitrate and phosphate from fertilisers and farming runoff contribute to eutrophication[3]. A spectacular example of this is the occurrence of "green tides". Pesticides, or phytosanitary products, can have toxic effects on aquatic species. Like nitrates, they sometimes make water treatment

[3] Eutrophication occurs when an environment becomes enriched with nutrients. This new abundance encourages the proliferation of species able to use these nutrients, to the detriment of native species. When the extra biomass that is produced begins to decompose, it consumes the oxygen in the aquatic environment and profoundly affects the role of this environment in the ecosystem.

for drinking water more costly. In the most developed countries, the reduction in urban and industrial pollution has increased the relative percentage of agricultural contaminants and their visibility. In France and elsewhere in Europe, the use of purification stations has reduced the quantity of organic matter that is released into waterways and thus limited oxygen depletion, which negatively affects the development of aquatic biodiversity. The role that nitrates and phosphates play in the eutrophication of waterways has become clearer and agriculture often plays a leading role in the spreading of these pollutants. Catchments used to collect drinking water are often found in rural areas and are affected most often by agricultural contaminants. The use of chemical inputs and the concentration of farming appear to be the main causes of diffuse pollution.

When irrigation is used to water large areas, it can lead to greater tensions than those caused by a deterioration in water quality. The majority of fresh water use is tied to irrigation. It accounts for around 70 % of fresh water use worldwide, but this figure may be as high as 90 % in irrigated areas. What is more, irrigation is inevitably used at times when water is scarce. Unlike other water uses, most of the water used for irrigation evaporates off the surface of leaves or the ground, and only a small amount returns to the source from which it was taken. Irrigation sustains the most productive forms of agriculture, which contribute to approximately 40 % of global production [3], even though they only represent 15 % of cultivated areas (i.e. just over 200 million hectares out of the 1,600 million hectares of currently cultivated land), but the water requirements of these areas were multiplied by approximately 2.5 between 1950 and 2000. In numerous regions, need has surpassed the limits of sustainable water use. Certain rivers wich appear "dried up" in France during periods of critical drought are examples of this, but the French phenomenon is nothing compared to the devastation of the Aral Sea. Water systems may also become affected when deltas or groundwater become salinated through contact with sea water. Groundwater levels are dwindling, which means this source is being pumped deeper and deeper and risks depletion, particularly the Ogallala aquifer in the central United States, which satisfies irrigation needs for a quarter of the country's agricultural land.

Irrigation can also cause serious problems with water quality, for instance when water was contaminated with arsenic in northern India, leading to serious health problems. If the water that is used to irrigate contains salts, even at low concentrations, it can cause soil degradation by increasing soil salinity when appropriate drainage techniques are not employed. This can have very serious effects on production, which may even lead to the abandonment of the land. A consequential proportion of irrigated surfaces is therefore in danger (half, according to certain authors) and 20–30 million hectares may be very affected [2]. The growth of competing needs (energy, industry and domestic life) has led to a growing intensity in conflicts of use in the many countries and regions that are affected by what is called "hydric stress", that is to say an insufficient availability of sustainable water per capita. Forecasts for demographics, economic growth and climate change predict increased tensions in sensitive zones, such as those near the Mediterranean. Finally, agriculture's destruction of natural plant cover, and practices that leave soils uncovered between crop rotations encourage runoff, leaving hillsides destabilised. Without protective structures, there is thus an increased risk of dangerously strong runoffs, catastrophic flooding or even landslides. These effects take on cataclysmic proportions after massive deforestation, such as occurs in Haiti or Bangladesh, for instance. With highly intensified agriculture, as exists in Western Europe (France, for instance), the expansion of fields favourable to an optimised use of machinery and the elimination of obstacles to runoff (hedges, banks, paths, etc.) have contributed to more frequent and more serious mudslides in important agricultural and wine-growing regions.

Agriculture's Impact on the Atmosphere and the Climate

The effects of agriculture on the atmosphere are a more recent concern than those on water, but they have become a priority for the coming decades. Direct agricultural emissions of GHGs make up 13.5 % of total global emissions [4], in terms of "global warming potential"[4,5]. Changes in land use, due mainly to converting natural ecosystems to farmland, are responsible for a further 17.4 % of emissions. Destruction of natural biomass—particularly burning—and losses in the soil's organic carbon after cultivation are the main causes of this. The expansion of agriculture is a major factor in the intensification of global warming. Excluding effects from land use changes and the carbon content of soils, the three main components of agricultural GHG emissions are:

- Nitrous oxide (N_2O), approximately 40 %: most emissions come from the microbial transformation of nitrogen in the soil, which is transferred there in mineral form (fertilisers) or organic form (livestock farming runoffs). Agriculture is responsible for more than 70 % of these emissions worldwide.
- Methane (CH_4), approximately 50 %: produced by ruminants, livestock waste and the soils of underwater rice fields. Half of global methane emissions are generated by agriculture, two thirds of which from livestock farming.
- Carbon dioxide (CO_2), less than 10 %: emissions come from direct energy consumption by farming machinery and heating.

However, this does not include indirect emissions due to the production of agricultural inputs and equipment. Nitrogen-based fertilisers are particularly important here because of the high energy cost of their production, the GHG emissions of which come to nearly half those that arise in spreading the fertiliser. International regulations for measuring emissions assign them to the industrial sector but it is quite clear that they are the effects of agriculture on global emissions. Synthetic

[4] In France, direct agricultural emissions are estimated at 21 % of the national total. They are second only to transportation-based emissions, which represent 25 %.

[5] The "global warming potential" (GWP) is a representation of the effect a gas has on the heat balance of the atmosphere. It combines the ability of a gas to absorb radiation and its atmospheric lifetime. It is calculated based on a lifetime of 100 years and expressed in relation to the GWP of CO_2. Methane and nitrous oxide have a GWP of 22 and 296 times that of CO_2, respectively.

nitrogen fertilisers thus appear to be the main cause of GHG emissions from intensive agriculture. Their use is also increasing more rapidly than other types of fertilisation, with a 22 % rise between 1990 and 2005, while total agricultural emissions only rose 17 %[6]. The issue of methane emissions from ruminants is more complex. The increasingly intensive farming of animals for meat and milk production, combined with a more concentrated diet, reduces the ratio of methane emitted to output generated, but the accumulation of organic carbon in permanent grassland soils may make up for this apparent advantage of intensification through carbon dioxide sequestration from the atmosphere, shifting the argument in favour of grass-fed farming.

The respective contribution of different livestock systems to methane emissions, and the link between their intensification and GHG emissions (taking into account emissions generated by feed production) have not yet been satisfactorily measured in order to form an educated opinion about the effects of intensive systems. From this point of view, the global incrimination of the impact of livestock following the FAO's report on the subject [5] led to favouring monogastric animals, which emit less methane and whose transformation yields are greater than those of ruminants. An evolution towards monogastric livestock farming would mean a shift in consumption towards "white meats" (i.e. pork and poultry), which account for the majority of the global increase in meat consumption. Whatever the outcome, these two key factors—emissions from mineral fertilisers and methane emissions from animals—in agriculture's impact on global warming represent a major issue in the evaluation of its environmental performances. Their importance in this debate will continue to grow in the coming decades, especially because increased yields have had several positive effects, helping to avoid land use conversion, reduce GHG emissions and preserve natural habitats through carbon sequestration.

Greenhouse gases are not the only agricultural emissions that affect the quality of the atmosphere. Ammonia emissions, the main source of which is animal waste, are also an issue. After being dissolved by the rain, ammonia enters the soil and raises its acidity level. These rains also bring nitrogen, which plays a role in the increased growth rate of European forests, that has occurred thanks to the fertilising effect of nitrogen. There are also pesticide emissions, which can be quite high during crop treatment periods. The aerial application of pesticides has expanded beyond local applications and now covers large distances, as shown by the presence of DDT in the Arctic. Some pesticides evaporate from the ground outside of application periods and cause constant, chronic levels in the atmosphere. As a consequence, a diffuse presence of low concentrations of pesticides can be observed in the environment. The ecotoxic effects of these emissions and any ensuing health consequences are not well known.

[6] This is not the case in France and the EU, where the use of nitrogen fertilisers decreased, contributing to an 8 % reduction in total agricultural emissions between 1990 and 2008.

Soil: An Endangered Resource

As we have already seen, soil degradation will be a determining factor in the planet's productive capacities, even if it is less visible than effects on water and the atmosphere. It is important to note that soil is a resource that renews itself very slowly, over millennia, via the natural process of pedogenesis. It can only be renewed artificially through very costly, localised methods such as when rehabilitating industrial fallow land or polluted sites, using fill materials. This means that losses, mostly from erosion, can only be replenished if the rate of loss is very slow. Agriculture is not the only cause. Expanding infrastructure, exploitation of subsurface resources and industrial pollution have all played a significant role in the degradation of soil, or in rendering it unfit—whether completely or partially—for cultivation. Agriculture has played its part in several areas of soil degradation, notably the following:

Water and Wind Erosion

Erosion is one of the most serious threats to soil quality because it represents a loss of earth that reduces the thickness of the soil and thus its ability to support and provide plants with nutrients. The average rate of erosion worldwide is estimated at 1 mm/year, while soil is formed from parent rock at a rate of only 0.1 mm/year [2]. The surface area affected is considerable, approximately 1 billion hectares. Agriculture and livestock farming have significantly intensified this phenomenon by destroying permanent ground cover (grass or woods), which protects soil from the mechanical aggression of water and wind. Overgrazing is the main cause of wind erosion (in northern China, for instance) and uncontrolled deforestation is the main cause of water erosion. Neither of these factors is specific to intensive farming. "Traditional" agriculture caused serious erosion problems in the past and continues to do so. When no preventive measures are taken, modern agriculture increases the risk of erosion in several ways:

- The practice of tilling, which leaves the surface structure excessively fragile and enables the wind to remove particles, crust to form and runoffs to occur.
- Reduced levels of organic matter, which leads to the degradation of the soil structure.
- Simplified crop rotation, which can leave a significant amount of soil surface exposed to the wind and rain, for instance, during the winter between rotations.
- Enlarged plots and the destruction of natural features (hedges, trenches, margins, etc.), which slow or channel the flow of water runoff.

The combination of these factors intensifies water erosion in industrial crop areas in Europe and in France, even on slight slopes. It is estimated that 16% of European soils are affected by water erosion, but the role each of these factors plays has not been explicitly quantified [6]. Effective erosion prevention techniques do exist. The progress of conservation techniques, which reduce tillage intensity and encourage vegetation and the use of harvest residue between crop rotations, has brought improvements. Now we must find a way to better quantify local risks in order to motivate people to use these techniques and to accept their collective nature, where endangered slopes and watersheds are concerned.

Physical Soil Degradation

Physical soil degradation is linked to erosion because the mechanical and hydric properties of soil determine the ease with which water or wind removes particles and the intensity of runoff. It includes soil compaction from farming vehicles and the deterioration of the soil structure. Both impact the circulation of water and wind, as well as root penetration. Soil structure degradation may occur due to inappropriate tillage practices. The extent of degradation is very dependent on the abundance of organic matter in the topsoil.

Reduced Levels of Organic Matter

Organic matter plays a key role in the makeup of soil aggregates and the ability of soil to resist the aggressive effects of water. Cultivation and tillage can cause the levels of organic matter to drop significantly. They decrease most rapidly during the first few years, but continue to decrease over the long term. Deeper ploughing, due to the mechanisation of the process, as well as reduced levels of organic matter and the cultivation of grasslands has led to significant reductions in organic matter in European and North American agriculture. These low levels are considered by the European Union as one of the main threats to its soils.

Soil Salinity

Soil salinity has been already mentioned as a serious consequence of irrigation with low-quality water.

Losses in Biodiversity

Soil is home to a wide variety of living organisms[7]. Plants depend on these organisms to transform elements into the minerals they need (in particular, via the nitrogen and phosphorous cycles) and to ensure the general physical and chemical functioning of the soil, such as the transformation of organic carbon from plants. Tillage and low levels of organic material take a toll on soil biodiversity, which can also be affected by certain toxic inputs, such as copper used to protect crops from fungal parasites. It is difficult to measure the consequences of biodiversity loss and for the time being no satisfactory estimate exists. While documentation is currently

[7] There are approximately 10,000 living species in 1 cm^3 of soil.

scarce, research on the topic is being actively carried out, particularly thanks to advances in molecular biology[8]. This brief overview of the state of affairs shows that, even if we could inventory the process of soil degradation and measure its impact, it would be impossible to establish the amount specifically linked to "intensive" production methods. These methods were established based on agronomics and the maintenance of soil "fertility" using fertilisers and other soil improvement techniques and helped improve the potential productivity of soils, which in some regions was limited by acid levels and mineral deficiencies.

For its part, traditional agriculture is often unable to sustainably compensate for the nutrients lost through harvests and water flow. Small though these losses may be, they have led to very serious problems of soil nutrient depletion. This was the case in numerous regions in France at the end of the nineteenth century. A modern example is traditional, shifting slash and burn cultivation, which is still practiced today in Africa. When land is not left fallow long enough for the natural environment to renew soil reserves, this cultivation method leads to dangerously impoverished soils. Intensification thus plays a double role. Its focus on maintaining soil fertility ignores the importance of cycles and causes soil degradation. This degradation is caused by the mechanisation and specialisation of farming, as well as the now-common separation between animal and plant production and irrigation with salted water. This is what is meant by the saying "agriculture has forgotten agronomy". The fragility of soil requires that all types of agriculture strive to preserve its properties and functions and protect them with appropriate management measures.

Biodiversity: Agriculture and the Sixth Extinction

The current rate of species extinction is estimated at 100–1,000 times the "normal" rate, which has led many to call this a sixth extinction, in reference to the five major extinctions throughout geological time. Agriculture has contributed to this extinction by replacing natural ecosystems with agroecosystems (see Chap. 6). These decrease the heterogeneity and diversity of habitats and are managed in such a way as to control the species that might compete with the plants grown or the animals raised. Agriculture thus has a tendency to reduce biodiversity, especially when it overtakes very rich environments, such as tropical forests. However, in the long run it can also stimulate a biodiversity that is adapted to agriculture. In Europe, non-intensive permanent grasslands and grassy grazinglands happen to be particularly rich agroecosystems.

[8] Until very recently, the study of soil microorganisms was limited to those that could be isolated and cultured, which represented only 1–10% of the species present. The ability to extract and sequence the DNA of the soil made it possible to surmount this obstacle and research the genes present in the entire living soil community and their functions. This significant advance was a huge stride in the understanding of soil ecology, which, for the most part, is still a "black box", but should be better managed in the long term.

Because of the numerous situations across the planet and the difficulty of quantifying biodiversity[9], it is not currently possible to measure the global impact agriculture has on biodiversity beyond the effect of the destruction of natural ecosystems. There has been important research and development work in methods of species observation, but it still falls short of meeting the current challenge. The continual reduction in the biodiversity of cultivated areas has been observed in developed and developing countries. It has an irreversible effect on the genetic diversity of domestic species. Intensive agriculture focuses on the use of a reduced number of plant and animal species, and on the most improved varieties and breeds. It is thus necessary to develop an efficient strategy of conservation of genetic heritage which may be endangered. The homogenisation of agricultural species also leads to a reduction in biodiversity in the wild, as with pollinating insect colonies. Wild biodiversity is also affected by another factor of the homogenisation of landscapes: the reduction in the proportion of natural or semi-natural spaces (hedges, field margins, permanent grasslands, trenches, ponds, wetlands, etc.). The improved productivity of soils—through fertilisation, corrections in acidity levels or drainage—all contribute to this negative trend by eliminating the species adapted to these environments. Finally, plant protection products also contribute to the reduction or extinction of wild populations, either directly through their toxicity or indirectly through the elimination of their food supply (weeds, for example). The specialisation of farms and production areas, the production of a smaller number of plant varieties and animal breeds, our ability to profoundly transform environments and the chemical control of plant and animal pests all play a role in intensive agriculture and together lead to impoverished biodiversity in agricultural areas.

Consequences and Risks for Agriculture and Other Activities

The main effect that diminished resources will have on agricultural production concerns long term viability and costs. Agriculture affects the ecosystem services from which it benefits when it alters soil quality, uses water sources in unsustainable ways and reduces biodiversity to the point where it no longer benefits from pest control, pollination and improved soil structure. We have tried to evaluate the impact that post-green revolution agriculture has on this, keeping in mind that traditional agriculture is also not generally sustainable in the long term.

It is also important to take note of the non-renewable resources consumed by agriculture. First, there are the mineral resources used in the production of fertilisers, such as phosphate and potassium. Estimates of international reserves (though quite uncertain, as we know, given their price fluctuations [7]) suggest that they are in danger of becoming quite rare by the end of this century, at least as far as

[9] Available indicators are relative to the number of species observed, which only includes certain living species (vertebrates, insects, vascular plants, etc.). It excludes the numerous organisms that are difficult to observe and quantify, such as those that live in the soil, as well as the genetic diversity of each species.

phosphorous is concerned. "Peak phosphorous" and "peak potassium"[10] are most likely close at hand. As there is not currently any possible substitute for these elements, which are indispensable to heightened production, intensive agriculture remains dependent on them. It must therefore maximise the efficiency of their use and prioritise recycling in order to minimise losses[11]. It then becomes necessary to discuss the issue of the resilience of intensive agriculture, or its capacity to maintain its functionalities when faced with a disturbance, and to regain them after the consequences of this disturbance have been felt. This notion can be applied to natural ecosystems affected by climatic hazards, to extreme events such as storms and fire and to biological invasions. The concept is closely tied to biodiversity, because a high level of biodiversity contributes to increasing an ecosystem's resilience, as numerous results show. The simplification and homogenisation of agroecosystems, particularly intensive systems, weaken resilience. Intensive systems are most vulnerable to attacks from pathogens or pests, thus requiring the use of more complicated, costly and polluting pesticides. Planting diversified crops with different vegetative cycles (sown in the autumn or spring) minimises the average impact of a temporary dry spell. The genetic diversity of crops may limit the dissemination of a parasite epidemic, but the determining factor of human actions and the means they use to mobilise the functioning of cultivated areas and correcting disturbances makes it difficult to show and give an overall estimate of the loss in resilience of agricultural systems due to intensification. Agriculture's effects on resources also lead to a reduced access to these resources and higher costs for their non-agricultural uses. Water needs for industry, energy and growing domestic uses—especially in emerging countries and countries with urban and demographic expansion—make the question of irrigation increasingly problematic in numerous regions of the world, such as Mediterranean countries, as well as China and India.

On a more general level, by favouring production, agriculture can alter other ecosystem services, such as the maintenance of our biodiversity heritage, recreational and aesthetic uses of the landscape and carbon fixation via the accumulation of organic matter in soil. We can thus consider the supply of plant and animal products as one service among many that agriculture provides and which may have economic value. From a local point of view, weighing these different options may justify a reduction in the volume of output in order to take advantage of another service whose value would compensate the loss in production. For example, in areas where mineral water and other drinking waters are sourced, water quality preservation is often favoured to the detriment of agriculture and economic compensations. However, these local decisions are made independently of a more universal vision

[10] These notions are derived from the notion of "peak oil", which corresponds to the time when accessible oil reserves will begin to diminish, progressively becoming depleted after a prolonged upswing fueled by increasing demand.

[11] Research is also underway to develop a means of using these scarce or unavailable resources entirely, by accelerating the transformation of minerals by the organisms in the rhizosphere (the part of the soil modified by the presence of roots). The recent discovery by INRA researchers of Myc factors, which are active in the symbiosis between roots and fungi, is a real advance in this direction.

of their impact on global public goods, or goods with a global dimension, through a necessary transfer of the production lost. Once we accept the long-term absence of agricultural surpluses and the need to limit the expansion of agricultural land, it becomes ever more important to analyse the planetary effects of policies that strive for equilibrium between agriculture's different functions.

An Integrated Evaluation of Services Provided and Used by Agriculture

In conclusion, the evaluation of different modes of agricultural production is a highly complex issue. This evaluation must be integrated, which means taking into account the quantity and quality of agricultural products obtained, the ecosystem services obtained or lost and the resources consumed. Environmental evaluation methods, such as the Life Cycle Assessment (LCA) indicator, pose some classic problems, including the need for reliable data and the ability to incorporate such heterogeneous effects as energy consumption, GHG emissions and effects on water and biodiversity. It is well known that the monetary evaluation of these factors is difficult, uncertain and thus frequently disputed, but there is also the other general problem of time, since this evaluation must be carried out over the long term. The need to update those future values is a controversial issue with decisive consequences.

The debate following the publication of the Stern report on the fight against global warming thoroughly illustrated this point. The questioning of biofuel production highlighted the difficulty of the evaluation of activities based on the use of these lands, or leading to changes in land use. How do environmental costs vary with changes in the balance between national production and imports from Brazil and Indonesia? We see that the many links between agriculture and the environment and the taking into account of the different scales of measurement make evaluation very complex, especially because these are complex systems being evaluated. Formulating the results of this evaluation is also problematic. For the evaluation of resource management in a given region, the measurement used is environmental impacts per hectare of cultivated land, whether it is to evaluate the amount of nitrate emitted into aquatic environments or GHG emissions into the atmosphere.

However, if we consider that agricultural production is indispensable—which is the case throughout the world—environmental costs must be considered in light of total output. If the preservation of the environment comes at the cost of significantly reduced yields, then the results can be distinctly different. A method of evaluation must thus be developed which favours efficiency, i.e. the yield of transforming resources (inputs, water, energy, etc.) and environmental impacts (for the atmosphere, biodiversity and soils) into agricultural products and ecosystem services. Outlining and implementing this method will require tremendous effort, but the risk of making serious decision errors, such as international policies based on incomplete forecasts or inadequate indicators, cannot be underestimated. Agriculture is part of the challenge to better understand planet Earth. Agronomy, and its many subdisciplines, is tasked with tackling this issue, working in close collaboration with Earth, environmental and social sciences.

While each technical element of agriculture must be taken into consideration in its own right, and ways to improve them found, it is no longer possible to develop an intensive, ecological agricultural system by focusing on these individual factors alone. In order to determine whether or not improving these techniques will lead to real progress, we must be able to anticipate the effects these innovations will have on the overall production system, in which they play a key role. It soon becomes clear that there is a need for a new level of organisation, a way to evaluate food systems which would go beyond mere production systems. The scope of the issue of production localisation, which includes land use and land use changes, thus becomes broader, encompassing trade and the toll that transport takes on the environment. Available results on this topic show that the environmental costs of production, especially in terms of GHG emissions, are far greater than those of transport[12], especially for long distance maritime transport. It is thus apparent that the differences in production conditions among the different areas of the world remain an important driver for action in favour of a more ecological form of agriculture.

Agriculture in a Changing Environment: Climatic Uncertainties

Although its contribution to global warming cannot be ignored, farming—due to its very dependence on climate—is also hard hit by climate change and will continue to be affected at an increasing rate in the decades and centuries to come. A number of factors may affect its production capacities and the risks to which it will be exposed, including rising temperatures, changing levels and patterns of precipitation, extreme weather events (which will likely become more frequent), rising sea levels and new distribution areas of pests and parasites. Increased concentrations of atmospheric CO_2 will boost the efficiency of photosynthesis in plants, which will partly compensate for losses in production expected in certain regions of the world. Gains and losses vary greatly with geography and crop species. The adaptations needed to react to these changes will vary just as much and depend heavily on the resilience of the agricultural systems and societies involved; they may include training, information and research, investment capacity and insurance mechanisms.

If there is one thing more difficult to develop with certainty than global climate models, it is the careful application of these models on a regional level and their use in tandem with vegetation models. However, with all due caution, here are some predicted trends [8]. Subtropical and intertropical zones will no doubt be the most affected regions; by 2080, yields could decrease by nearly 50 % in the southern United States and Mexico, the Sahel and the zone spanning Lebanon, India, Cambodia and northern Australia. However, production in Russia, the northern United States and Canada, Europe and China could increase upwards of 35 %. Table 8.1 summarises this information.

These figures are admittedly very uncertain, but we cannot ignore the significance of the impact they predict. It will be necessary in the future to increase

[12] Fresh products transported via airplane are not analysed here. While the impact of this type of transport is a legitimate concern, it only accounts for a minimal volume of products transported.

Table 8.1. The effect of climate change on agricultural productivity by 2080: % change with (1) and without (2) accounting for increased concentrations of atmospheric CO_2. (Source [8])

	(1)	(2)
World	−15.9	−3.2
Industrialised countries	−6.3	+7.7
Developing countries, including	−21.0	−9.1
Sub-Saharan Africa	−27.5	−16.6
Middle East and North Africa	−21.2	−9.4
Asia	−19.3	−7.2
South America	−24.3	−12.9

productivity while compensating for the overall negative effect of climate change. The agricultural systems of developed countries will also have to step up their contribution to global food stocks in order to compensate for the growing deficits expected in the regions of the world confronted with rapidly growing food demand and more severe climate constraints. In the nearer future, particularly by the mid-twenty-first century, the trends described by different models become less clear. While the effects of rising temperatures and more frequent droughts could reduce overall yields, increased concentrations of atmospheric CO_2 could compensate for these losses, or even tip the balance in the other direction. However, experts debate the extent that this regulating effect could have [9]. As pronounced as the overall trend is, the question of variability in production, in conjunction with climate change risks, will reinforce the importance of the climate issue.

The link between climate change and the (statistically increasing) frequency of catastrophic weather events that destroy harvests (storms, hurricanes, floods, etc.) remains controversial. However, it is likely that the yields of certain crops will fluctuate dramatically when climatic thresholds of physiological stress are reached. These situations may include droughts, exacerbated by changing precipitation cycles; rising temperatures, which affect growth and development; or, at the other end of the spectrum, frosts affecting crops that had adapted to higher average temperatures [10]. Greater vulnerability of certain crops should be prevented through adaptation. Vulnerability includes the risks associated with epidemics that affect plants and livestock and which have spread with the expansion of environmental conditions favourable to disease-spreading agents. It also includes pests and, more generally, different types of invasive species, such as weeds. Specialists have identified a wide variety of these types of threats, and the effect of global changes on the migration of biological agents that are harmful to agriculture has been well documented in the past few decades[13]. It should be possible to adapt health protection practices in order to stem this tide quickly and effectively, and to put preventive measures in place.

If the only efficient option available is increasing treatment with toxic products, the ability to anticipate changes will be a key factor in avoiding having to

[13] A classic example of this method of dissemination is the western corn rootworm, which originated in the United States, but has spread to several regions in France, after first being observed at airports.

choose between crop protection and environmental preservation. Climate change will modify productivity, but it will also affect the relationship between agriculture and the resources it exploits, such as water and soil. Droughts will cause yields to decrease, but will also lead to a reduction in the volume of water used to irrigate renewable resources and a growing number of other competing uses. There could be serious repercussions on a local level, in terms of river flows and groundwater renewal, as the proportion of rainfall that trickles down beneath the surface decreases. The intensity of evaporation may also be reinforced by global warming, for example because of increased vegetation during the winter in temperate zones. These hydrological consequences will have an important impact on the outlook for irrigated land, which in turn plays an important role in the outlook for increased global agricultural production.

The progressive melting of mountain glaciers could, however, help to sustain river flows during a transitional period, which would be more or less sustainable, depending on whether this pertains to the glaciers of the Alps or the Himalayas. The gradual depletion of these freshwater reserves and decreased snowfall will contribute to increasingly pronounced low water levels during the dry season. Changing rainfall patterns will affect its seasonal distribution, for example increased rainfall during the winter with infrequent rain during the summer in temperate regions, or an increase in the frequency of very heavy rainfalls. These changes will have several consequences for soil. The risk of water erosion will increase, as will that of soil saturation or submersion and tilling may become difficult. Structural degradations may become more serious and N_2O emissions increase. The anticipated rise in sea levels will cause low-lying land in danger of submersion to progressively become unusable. Deltas will be particularly vulnerable. If sea levels rise by 1 m, up to 6 million hectares of land could be lost in Bangladesh and Vietnam [11]. This figure is much higher than the current predictions considered probable for 2050, but given the general lack of understanding of the functioning of polar ice caps and the uncertain nature of long-term forecasts, we cannot ignore the possibility of additional losses in arable land. Ultimately, the intensification of agriculture may well be a great success in terms of product supply and efficient use of cultivated areas, but it would come at the price of profound environmental alterations and an elevated consumption of rare and/or non-renewable resources. It will thus be necessary to develop and move toward new systems of productive, ecological agriculture, better adapted to climate change.

Increased, Improved Production: A Realistic Goal?

Towards a New Agricultural Revolution: "Ecologically Intensive" Agriculture

The current growth rate in land productivity must be maintained, but this is no longer possible without incorporating a method of managing the potential environmental impacts and a method of efficient resource use. What is more, the growing value attributed to ecosystem services (see Chap. 6) implies that the notion of "land production" is no longer limited to harvests and animal production but necessarily includes services. This notion must also aim to simultaneously use and enrich all the ecosystem services available in a virtuous circle with a positive impact on the production of agricultural goods. This is what is meant by "ecologically intensive" agriculture[14]. However, it is our ability to turn these concepts and goals into concrete actions that will be key for the development of a clear vision of agriculture in the coming decades and the changes required in order to achieve this. Around the world, agricultural situations differ according to the state of intensification as well as the productive performances already achieved by different agricultural systems and farmers from different parts of the world. The precise analysis of this diversity is a complex task. It should not be limited to a simple comparison of yields per hectare, but make reference to the potential level of production per unit, which takes into account physical factors (climate and soil), rainfed agriculture conditions and—when water is available—irrigation possibilities. The most efficient agricultures solely in terms of productivity per hectare are those which come the closest to reaching their full potential[15], using the most efficient genotypes, controlling the state of the soil and plant nutrients, using plant protection products and, where applicable, irrigation. Thus the difference between Australian and French wheat yields (an average of 1,650 and 7,130 kg per hectare, respectively, for the 2000–2009 period[16]) is not a reflection of different levels of efficiency in the production processes, but of the limiting role that rainfall plays in Australia's grain belt.

This standard of analysis leads us to distinguish the agricultural systems that are the closest to achieving their potential: as a gross oversimplification, they are the systems of Western Europe, North Africa, a large part of Asia and certain regions in Brazil and Argentina. The agricultural systems in these areas have exploited all the techniques and methods of the first green revolution, with the associated costs and impacts, adapting them to their local needs. Because of this, they are now all faced with the same twofold challenge. First, maintaining production growth, and the continual innovation that this implies, will become all the more difficult as production levels reach the physical potential of the local environment. Second, the imperatives of efficient resource use and reduced environmental impact will mean that innovative systems and techniques will have to be optimised, or even culled,

[14] The term doubly green revolution is also used (Conway G. 1998. The Doubly Green Revolution: Food for All in the Twenty-First Century. Ithaca, New York, Cornell University Press, Comstock Pub. Associates, 334 pages), though the two notions differ. Ecological intensification is more ambitious because it aims for a productive system of agriculture through the use of ecological mechanisms, while the doubly green revolution strives to take ecological services into account when measuring productivity.

[15] This potential depends on the amount of photosynthetically active radiation (PAR), the thermal regime and the water regime, determined by precipitation and the soil's water reserves.

[16] According to Australian Grain magazine (www.ausgrain.com.au) and the French General Association of Wheat Producers (AGPB—Association générale des producteurs de blé), respectively.

based on different performance criteria, possibly to the detriment of the criterion of production. We will illustrate this aspect using the current situation in France, which is representative of this type of problem. Besides these technical aspects, we must also take into consideration different social aspects, not only for professionals in the agricultural sector (farmers, for instance), but also for the millions of men and women who depend on the sector. Insert 8.1 and Fig. 8.1 explore this issue.

Insert 8.1. What Do OECD Agriculture (Taking French Agriculture as an Example) and Asian Agriculture have in Common?

Both systems of agriculture produce approximately 25,000 kcal/day/ha on average. In this regard, they are leaders worldwide. Their difference lies in the way they reach this level of partial productivity of land. Both make use of intermediate consumption, but to differing degrees. While French agriculture achieves an exceptionally productive agricultural output of 420,000 kcal/day/labourer, Asian agriculture only produces 10,000—or more than 40 times less. Ecological intensification can thus not be carried out in the same manner in Beauce (a fertile region south of Paris) as in the Mekong Delta; the same goes for social and macroeconomic consequences. In other words, the determining factors for agriculture will not only be technical, but also societal. Thanks to the vast diversity of world regions, several agricultural models coexist, representing different social models.

Other agricultural systems have put into practice very few of the green revolution technologies, and their production performances are still far from reaching their full potential. This is true for countries whose insufficient development and prolonged lack of investment in agriculture have prevented local producers from having adequate access to production factors such as improved seeds, water, fertilisers, phytosanitary products, tools and machines. These countries include several African countries, especially in sub-Saharan Africa. It is important to note that the development of these agricultural systems should now be guided by the principles of ecological intensification, without stopping at the green revolution stage first and reproducing the mistakes analysed above. These countries are not the only ones experiencing "under-production". Countries whose development and capital allowed them, at least temporarily, to take part in the green revolution, but whose socio-economic organisation and insufficient investments impeded its success, such as countries under the influence of the former Soviet Union, also fall in this category. This is an important region in terms of surface area and the potential for production; Ukraine is a striking example.

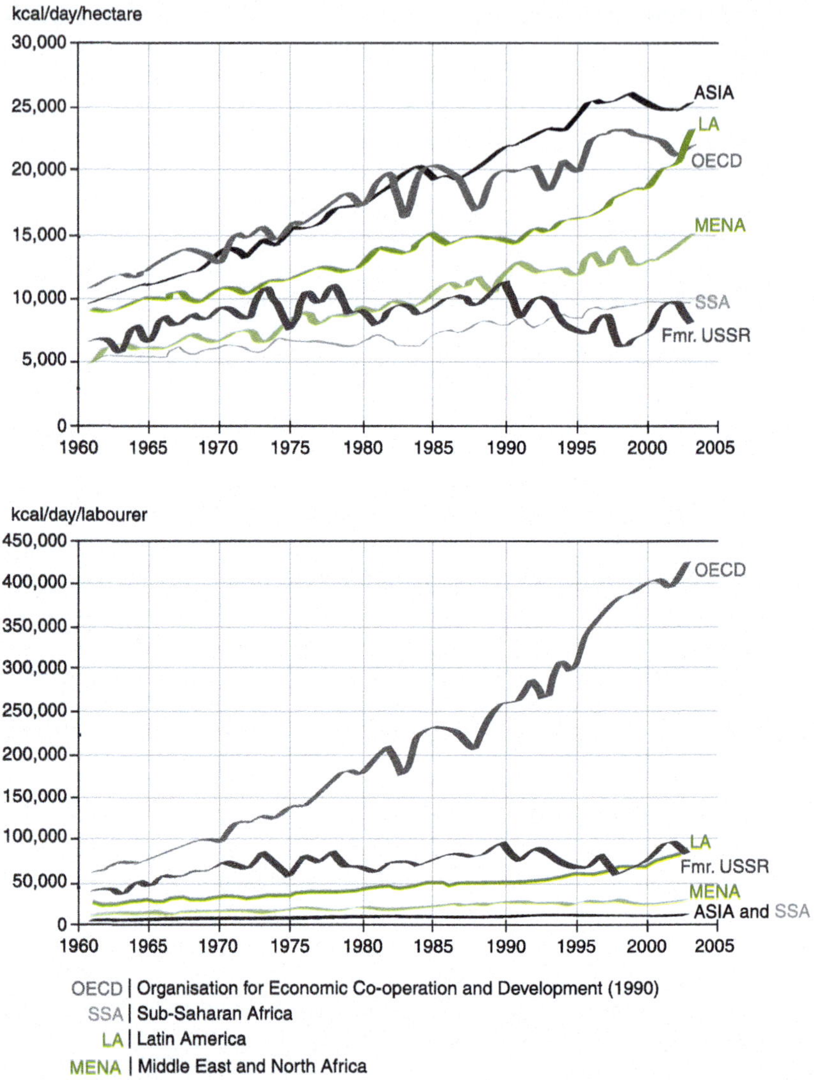

Fig. 8.1 Panel a: Food crop production per hectare of cultivated agricultural land (1961–2003). Panel b: Food crop production per agricultural labourer (1961–2003). (Reproduced with the permission of F. Bourin Publishers)

Challenges Facing Intensive Agriculture: The Example of French Agriculture

France's agricultural situation illustrates the complexity of the challenge facing highly productive agriculture (by hectare and by labour unit). Even if the volume of French production is not sufficient to play a key role in the changes expected to occur in global supply, we cannot ignore the global consequences of the future of its agriculture, especially considering the flows of French exports and imports, and their impact on other productive regions involved in these exchanges. Once we accept this way of thinking, we can consider socio-economic issues (which are mostly tied to production volumes) and environmental goals in another light, by weighing the environmental value of the production volume, including negative effects avoided in other areas.

Tension Between Production Levels and Environmental Objectives

This tension arises from the additional costs and possible yield losses associated with implementing measures to preserve or restore environmental goods and services. These measures reflect different technical principles and levels of action that can influence production.

Controlling Environmental Impact

The main objective, mostly concerning aquatic pollution, was the development of techniques to limit and, if possible, eliminate the spread of pollutants, without modifying the production process. To fight against the spread of nitrogen and pesticides through winter infiltration and water runoff and protect winter soils, a living or dead plant cover was used and grass buffer strips were established along waterways. These solutions were developed and promoted through public policies. Technical expertise is generally sufficient in controlling most of the possible negative effects on production, with the noticeable exception of grass strips, the construction of which reduce agricultural output. However, on their own, these measures are not efficient enough to achieve the environmental quality standards for aquatic environments, as specified in the EU's Water Framework Directive. Beyond limiting the spread of toxic chemicals, these techniques can have positive effects on protecting soils against erosion, improving their structure and their agronomic maintenance, particularly when used in conjunction with "conservation" practices which involve the abandonment of ploughing. However, where the environment is considered, benefits are still only partial. Even though doing away with ploughing leads to a significant reduction in fuel, more herbicides may be required. In extreme cases, these "direct sowing" techniques, where tillage only affects the sowing line, can lead to the contamination of groundwater by herbicides. They also appear to be statistically linked to yield losses of approximately 0.3 t/ha for French wheat [12].

Reducing Environmental Pressures

In order to reduce pollution more effectively, it will also be necessary to reduce the use of potentially polluting substances, such as pesticides and mineral or organic

Table 8.2 Expected effects of reducing pesticides in French field crops: Treatment Frequency Index (TFI)[a] and production value (with 100 corresponding to the national average for 2006). Adapted from [13] and [14]

	TFI	Production value	Gross profit margin (€/ha)
Intensive protection	142	105	455
Supervised protection	106	103	498
Integrated protection	64	94	480
Integrated production	47	88	460
Organic production	–	64	341

[a] The Treatment Frequency Index (TFI) measures pesticide use. It is calculated for a given plot of land and expresses the ratio of the dose of the active substance applied to the recommended dose, for all treatments

fertilisers. Used in conjunction with the environmental impact controls discussed above, this should help reach the environmental objective concerning contamination levels of protected environments. In addition, the reduced use of chemical and energy inputs may have a positive overall effect on the amount of resources used to produce them and the GHG emissions that this production implies. However, when it comes to plant production factors such as fertilisers and protection against pests and diseases, reducing their use means increasing the risk of harvest losses. This risk must thus be compensated for through the use of alternative or preventive techniques, the efficiency of which will determine the cost associated with this reduced dependence on agricultural inputs. As an example in the specific case of phytosanitary products, INRA's "Ecophyto R&D" study [13] analysed the different scenarios of reduced pesticide use in mainland France, using experimental data and available statistics. Table 8.2 summarises this analysis for the case of field crops of cereals, oilseeds, sugar beets and potatoes.

Supervised protection: correcting "inefficiencies" As the table shows, the move from an intensive strategy of phytosanitary protection, which systematically prevents all development of plant pests, to a "supervised" protection strategy takes place without a significant loss of production and has economic benefits for the farmer (by saving on costs while maintaining output). After a more rigorous evaluation of the resources needed to obtain desired objectives of protection and production (such as ever-more efficient decision support tools), we see that this is due to a correction of technical inefficiencies. Fertilisers may also be used more efficiently: tools for calculating the dosage and schedule of application help to reduce excesses which constitute pollution risks, while maintaining an objective of high production adapted more strictly to its local potential. Decision support tools may be refined using sensors, which can give real-time information on the state of vegetation and its variability within the plot. When these sensors are located on satellites, we talk about precision farming. Precision farming leads to additional savings of inputs and benefits the farmer, who can make real use of these signal machines. As long as costs can be amortised, which is the case for large farms with large plots or when tools are shared among several smaller farms, improving upon this supervised strategy will lead to consequential progress.

Integrated protection: using alternative and preventive techniques In order to use chemical treatments less frequently without excessively increasing the risk of attack from pests and damage to crops, it becomes necessary to modify other practices. For field crops, this may include using varieties that are more resistant to parasites or modifying the schedule and density of sowing in order to discourage the multiplication of plant pests. Table 8.2 shows that this integrated protection strategy is equated with a modest reduction in productivity levels. However, it also negatively affects the economic margin, at least at the price levels for the products and inputs from the reference year of the study, which is 2006. The result is similar for other crops—vines, fruits and vegetables—for which other techniques may be used: for example, biological control of insect pests or installing protective nets. This shows that given our current state of knowledge and technology, chemical molecules are more effective in controlling pests than alternative methods, which are only partially effective. Multiple methods must be used together and be managed in such a way that they are precisely adapted to the health conditions of the plot in order to achieve a satisfactory result. They are thus more demanding of agronomic expertise and observation from farmers.

Integrated production: rethinking cropping systems If we are to continue toward reducing synthetic phytosanitary products, we must not only practice integrated protection on all the crops of a crop rotation, but also modify the crop rotation itself. The traditional strategy is—or was—to limit the populations of parasites or weeds best adapted to one or several of the crops present for too long or used too often on a plot by introducing crops that discouraged them. The use of pesticides made this process unnecessary and helped to simplify crop rotations, which in turn reduced the number of species cultivated and benefited the most profitable ones. If we wish to return to this pre-pesticide method, it will be necessary to diversify the rotations and make them longer by introducing less economically profitable crops. This explains the loss of the high-end production value in Table 8.2: it is associated with the development of crops such as peas, lucerne or triticale, when we move toward integrated production. At this price, integrated production makes it possible to respect the commitment of the French Environment Round Table (*Grenelle de l'environnement*) to reduce pesticide use by half by 2018, "if possible". In the extreme scenario, the generalised practice of organic agriculture, which forbids the use of synthetic pesticides, would lead to an approximately 40 % drop in production according to national statistics, as well as considerable decreases in agricultural margins and revenues, assuming these losses were not compensated for through substantially higher sale prices of agricultural products.

In short, the Ecophyto R&D study shows that our current knowledge and expertise will not be enough to maintain current levels of production while massively reducing pesticide use beyond the correction of "first-line" excesses linked to the apparently inefficient use of pesticides. However, these "first-line" excesses are far from negligible. There is a real tension between production goals and environmental issues. This is also true for the use of mineral fertilisers. Fertilisation gives plants the nutrients they need to grow, and so it must be strictly adapted to the yield objectives of the farmer. Available nutrient reserves in the soil can only contribute to higher yields in a limited way, and only temporarily if these reserves are not renewed through external sources. Organic fertilisation, which mainly consists of livestock effluent, is a resource which can, at least partly, be substituted for mineral fertilisers. But this means a displacement of fertilisation from the use of mineral fertilisers or the exploitation of soil reserves in other areas, at least for phosphorous and potassium. This is not the case for nitrogen, thanks to the alternative of using nitrogen-fixing plants such as legumes. The surface area used to cultivate these species can thus penalise plant production when their only possible use is for animal feed (as is currently the case for clover and lucerne), and when the price situation makes it more profitable for the livestock farmer to use other, protein-rich raw materials (e.g. oilseed cakes). In France, compensating for the disuse of nitrogen fertilisers by cultivating lucerne—the most efficient nitrogen-fixing plant that can be cultivated in France—would mean dedicating several million hectares to this crop. From this point of view, importing soybean, a legume used for animal feed, constitutes a transfer of nitrogen and, as a result, a source of synthetic fertiliser savings by recycling animal excrements.

Insert 8.2. Organic Agriculture and Global Food Security

When measured in hectares, organic agriculture has environmental benefits on several levels, including reduced pesticide residues and fertiliser pollution, improved biodiversity conservation and reduced GHG emissions. But is it realistic to develop this type of production on a large scale, or even a global scale, when land is becoming scarce and agricultural production must be increased? Excluding synthetic inputs (e.g. fertilisers, pesticides), organic agriculture produces less per hectare than so-called intensive production: for example, in France, it produces 10–40 % less per hectare for grain maize and 70–80 % less for peas. Given the current state of affairs, especially in terms of scientific and technical knowledge, organic agriculture cannot take the place of intensive agriculture in developed countries without a considerable decrease in output. However, in many developing countries, for example in sub-Sahelian Africa, where yields are lower and technical and financial resources less available, the adoption of organic agriculture would not cause this kind of drop in production.

However, on a global level, it turns out that environmental benefits measured per hectare are more than cancelled out when the issue is considered in terms of output, or the increase in global agricultural production needed in order to feed 9 billion people in 2050. To produce 1 t of organic wheat, more hectares are needed than for standard wheat; to raise an organic chicken for 100 days, more water and food is needed than raising a standard chicken for 45 days. The environmental benefits per hectare for organic agriculture decrease further when we account for the fact that the production deficit of this hectare must be compensated for by increasing production "elsewhere", thus through a change in soil use, which could have serious environmental effects. The second argument for organic agriculture is health benefits. For the farmers themselves, the risks linked to the use of toxic chemical substances are reduced as much as possible, even if appropriate precautionary measures used when handling treatment products already provide some protection. But it is clear that this is far from the case for all farms throughout the world. As for consumer health, caution must be exercised: many studies on the subject have shown a decrease in pesticide levels; however, results on the benefits of other composition changes, such as polyphenol levels in fruits and vegetables, are inconclusive or contradictory.

The fact remains that organic agriculture is an innovative production method because of the constraints it imposes on itself and its continual search for solutions that optimise the functioning of agro-ecosystem services, including legumes in plots, cropping plans and rotations, diversification of species, natural regulation of pests, etc. These are all rich learning experiences for other types of agriculture that are more ecological than intensive agriculture, such as systems of integrated protection or integrated production. Organic agriculture is also often highlighted for its participative approach which brings together researchers, experts from the development sector, farmers and consumers. There is much to learn from these experiences of cooperatively constructing a "new agriculture".

Sources: Guyomard 2009 [15]; Prospective France 2025 [16]; INRA Ecophyto R&D Study 2010 [13]; Doré T., Maraux F. 2010 [17]; Meynard J. M. 2009 [18]; FAO [19].

Managing Cultivated Areas and Landscapes

The third method of improving the relationship between agriculture and the environment puts into place actions that go beyond plots and farming. It concerns territories, on all different levels. The importance of determining the level pertinent to the ecological process at hand (see Chap. 6) is the main reason territories must be taken into account. It is clear that the conservation of water resources is only effectively dealt with at the level of watersheds, or the supply area for groundwater. Not all the areas of the same watershed present the same pollution risks. What counts is the accumulation of the different pollutants, not the one of which each plot is independently the source. In the same way, maintaining biodiversity depends on the existence of different landscape components with different functions: refuge and habitat, food resources, etc. Thus, maintaining beneficial organism populations capable of limiting the quantity of crop pests is positively correlated with a sufficient proportion of forest area, grassy zones or hedgerows in the landscape. Crop practices normally favourable to biodiversity have little effect if they are carried out in a uniform agricultural area, specialised in a small number of crops and devoid of semi-natural areas. This means we must picture the diversification of the landscape on a scale adapted to the endangered biodiversity. This notion is also important for the management of the genetic diversity of cultivated plants. A plant's genetic resistance to parasites is quickly overcome through parasitic adaptation when the use of resistant varieties becomes widespread. The presence of non-resistant varieties in the agricultural landscape can significantly slow down this process, but the expansion of semi-natural areas affects output, even if the decrease in production is only partial, as is the case, for example, for permanent grasslands. The production of ecosystem services has a trade-off, even if some of these services also benefit agriculture, such as the maintenance of pollinating populations. This is also true for the diversification of crops, as we have seen in terms of reduced pesticide use.

How do We Overcome the Trade-off Between Agricultural Production and the Supply of Ecosystem Services?

This description of the current state of affairs should not lead us to renounce the idea that agriculture can be both beneficial for the environment and more productive. New, effective practices can now be applied, such as crops that associate several varieties of the same species, but with different parasite-resistant genes. The obstacles to overcome in this area are intertwined with the supply chain: seed suppliers, storage firms and product processing. Diversifying their product ranges is complicated and costly (for example, multiplying the supply chain or adapting the processing methods). Other options, such as the simultaneous cultivation of different species in the same plot (for example one cereal and one legume) which lead to savings in nitrogen fertilisers and pesticides by reducing the risk of parasites, must still be improved in order to be implemented on a large scale on farms. In particular, we must first better understand the mechanisms that encourage both environmental and agronomic benefits. Among these mechanisms, facilitation (one species encourages the development of another, either by transferring nutrients or by preventing parasitic attacks) and complementarity (species use different resources for their growth, as is the case for species with different root depths) merit further study. It will also be useful to specify rules for choosing associated species and varieties, and to work on methods of cultivation that allow for maximum benefits. Despite the inherent difficulties, these areas of study are promising for both environmental and agronomic progress. Improving upon results obtained by a system which has been optimising production for 60 years, tied directly to efficiency and low-cost industrial inputs, will be difficult. In order to achieve this, it will be necessary to develop biotechnical and organisational innovations which will be further stimulated by well-defined public policies which provide incentives. It will be necessary to choose targets for incentive measures in accordance with a strict analysis of costs and benefits, integrating both the private, for-profit dimension and the public, non-profit dimension.

The Role of Genetically-improved Plants

Other types of innovations could modify this diagnosis and accelerate the rate of progress, both productive and ecological. One example is the genetic improvement of plants. In the past few decades, this has essentially been oriented toward researching the plants best adapted to prevailing cultivation conditions, ensuring near-optimal conditions in terms of minerals and health protection[17]. Improving the biochemical composition of harvested products (protein and oil content, etc.) has also been a priority for selection and, it must not be forgotten, one that was quite successful. The search for genotypes adapted to other crop systems, which pose disease and plant pest risks or limiting physical or chemical conditions, may help overcome these handicaps. Similarly, focusing on species overlooked by selection, due to the lack of a sufficient market for them, may help compensate for their economic handicap, by increasing their yields. One such progress is the availability of new possibilities of diversification and modification of crop rotations, for example by encouraging a new expansion of legume crops. In this sense, public support policies, even temporary ones, are justified by acting as a support tool for an "emerging activity". This is why INRA dedicates considerable research resources to improving peas, in order to improve their resistance to the cold and to increase their yields, and thus increase their competitiveness which is currently insufficient, and which therefore explains the significant reduction in surface areas dedicated to pea production. Stepping up investments in terms of the selection of cultivated plants used as winter ground cover, before the spring crop or as part of a mixed plant cover, may increase the environmental benefits and/or the production capacities of these kinds of dual-species covers.

Genetically-improved plants thus emerge as a key to increasing the agricultural and environmental performance of a productive and ecological agriculture. There are two major obstacles to overcome so that improved plants may meet these expectations. The first is to clearly define the selection criteria. This will mean anticipating the conditions of the crops planned, both in terms of crop system practices and climate risks. Determining the desired traits of the plants of the future is an important issue to research in its own right. This is why it is important to encourage modelling work which uses computer-aided tools and incorporates input from specialists in biology, genetics, agronomy, plant protection and bioclimatology. It is also important to encourage research which draws upon the economic and social sciences, with a view to shedding light on possible future scenarios for market conditions as well as social and regulatory conditions in which these will operate. The second obstacle is to reduce selection costs, shorten the length of the selection cycle and take into consideration the increased number of traits for which to select. Cost is the main reason that genetic improvement efforts currently focus on a small number of species, the generalised use of which would ensure amortisation. Cost is thus the main obstacle to diversification. Reducing costs will be a deciding factor in the progress of selection methods expected to be made thanks to advances in

[17] However, parasitic resistance was one of the main criteria for selection. Complete, single gene resistance to a disease was given priority. When managed in a non-sustainable way, these resistances are easily overcome through genetic adaptation of parasites which render them ineffective. Cumulated partial resistance may prove to be more sustainable.

genomics. Genome sequencing, initially carried out on species with single-stranded genomes (such as model species, but also rice), but now extended to other major agronomic species (maize, wheat and tomatoes) makes it possible to determine the position of genes and to identify those responsible for the variations of the traits to be improved. It is thus possible to perform selection based on molecular markers, which have been validated by studies under controlled conditions or in the field. We can reasonably assume that once it becomes possible to perform this kind of selection (what is called genomic selection), using these markers which are located throughout the genome, the rate of genetic progress should double.

Insert 8.3. Ensuring Long-term Resistance Through a Combination of Genetic Advances, Improved Varieties and Agronomic Innovations at Different Levels

Genetic resistance to disease or pests may be either qualitative and thus controlled by only a few genes (for example, the apple tree's resistance to apple scab) or quantitative and thus controlled by many different genes (for example, rapeseed's resistance to collar rot). This second type of resistance is usually less easily circumvented. However, as a general rule, if a variety which is resistant to a certain disease or pest becomes more common, there is a strong chance that its resistance will be circumvented: strains of disease or breeds of pests will adapt to this resistance and stamp out the work done to select for and identify this resistance and implant it in a variety with high agronomic value. The same thing happens if a group of varieties with the same resistant genes is cultivated on a large scale. It has become necessary to implement production methods that take into account the risk of resistance circumvention and allow for genetic progress to be of long-term value. Many drivers exist to accomplish this, from the crop scale to the landscape scale. Prolonged crop rotation cycles and the introduction of diversified species (such as legumes in the crop rotations of grains and oilseeds) will necessarily limit the proportion of one species in a given territory, which will reduce the likelihood of disease to affect this species. The creation of mosaic landscapes, with reasonably-sized plots, will also contribute to the reduction of epidemics.

At the scale of a group of plots, using well-chosen varieties with different resistant genes also helps to limit the development of epidemics. Working on plot margins may also have a positive impact. This was demonstrated by Petit et al. [20] to be an effective method of managing weeds in field crops, as well as in orchards, where the choice of species planted must be made based on their ability to host the same pests early and so stimulate the development of beneficial organism populations [21]. Crop practices will also play an important role. For annual crops, modifying sowing dates may help limit attacks by certain fungi and pests. Modifying the sowing rates and nitrogen fertilisation in the case of annual plants—or pruning methods for orchards, by modifying the foliage density in the plant cover—also helps reduce parasitic attacks. Long-term resistance management is a key issue that requires continuous research in the biology of plant species and the disease and pests that affect them, as well as training users and advisors in order to put drivers for action to good use. Pathogens, pests and diseases adapt. Confronting this constant adaptation requires the simultaneous manipulation of all available drivers, from genetics to crop systems, from improved varieties to crop associations in fields. Constant invention and innovation is an essential tool for developing the agriculture of tomorrow.

If costs can be reduced enough, then it will be possible to adapt plants to a diverse range of production conditions, rather than modifying environments to be best adapted to a standard genotype. The enormous strides that have been made in the productivity of genome sequencing may prove these hopes to be justified in this sense, but on the other hand, the investment that will be necessary in order to use these findings could lead to increased focus on a smaller number of targets. Progress in genetics will greatly depend on the resources dedicated to research, which will in turn depend on the presumed profitability of investment. This profitability is determined by the structure of varieties and by the type of protection used. In sum, genetic improvement will strongly determine the production gains of intensive agriculture in the future, whether in terms of output, quality or the conservation and optimum use of ecosystem services. This goal gives great significance to the availability of genetic resources, which constitute an important tool in the use of variability in the selection process. Preserving them, managing them and establishing laws that recognise them will determine what we can expect of future progress. Their cost and the unbalanced distribution of these resources worldwide make it necessary to share the task on an international level.

Beyond Varietal Selection: Other Biotechnological Solutions

The contribution of biology to agriculture is not limited to the genetic improvement of plants. Improved knowledge of the biology of soils, bacteria or symbiotic fungi and interactions between plants such as allelopathy[18] may lead to the development of new products or new methods of improving production, replacing chemical inputs or reducing environmental impacts. Research in biocontrol methods for plant diseases, which may take the form of molecules or natural organisms, is very active, as is research in the causes of their resistance to diseases. However, these methods are currently only partially successful and require further development and verification. Rapid advances in our knowledge of the genetic and physiological

[18] Allelopathy refers to the biochemical mechanisms of one plant species which allow it to inhibit the development of another. Research in this field is very active, notably in the area of weed control for crops.

mechanisms at work should allow us to overcome this roadblock and to remove uncertainties. INRA has used the results of analysing cell mechanisms of interactions between plants and symbiotic, nitrogen-fixing bacteria, and identified a biochemical factor of these natural interactions (the "Nod" factor). The molecule involved is used in the field in order to increase the productivity of soybean crops by improving their nitrogen fixation. Improved knowledge of soil microbiology will open new horizons in the long term, such as being able to influence the nitrogen cycle, in order to limit nitrous oxide emissions, for example. A firm commitment to an agriculture that uses chemical inputs more efficiently would encourage investment in this type of research, by ensuring they are valued by the market. And, once again, public policies can play a decisive role, by initiating a virtuous cycle.

Insert 8.4. A Revolution in Selection Methods

Selection methods are undergoing a veritable revolution, both in terms of the characterisation of phenotypes and our increasingly precise knowledge of genotypes.

Phenotyping and genotyping Phenotyping is the description of genotypes in a given environment and was traditionally conducted in the field or under controlled conditions (greenhouses and growing rooms). In the field, it is possible to analyse a large number of individual plants, but with less precision as to the environment of the plant to be characterised. Under controlled conditions, the characterisation of the environment is better, but the number of plants observed is limited by the available space. The situation is beginning to evolve significantly with the emergence of high-throughput phenotyping [23]. It is now possible to precisely and continuously characterise the growth of different features, such as roots [24] or fruit quality. It is also possible to more precisely characterise the environment of plants in the field. These data are useful inputs for growth models and help to better determine the genetic components of observable phenomena.

The development of molecular markers which are more evenly spread throughout the entire genome will make it possible to construct genetic maps and to localise the genetic factors responsible for traits such as shape, chemical composition or resistance to disease. At the same time, genetic sequencing and annotation make it possible to catalogue all genes and their presumed functions. Combining these two sets of data will enable the identification of "candidate" genes, which may be responsible for the desired function. Because genetic sequencing was not initially carried out on plants, it has been possible to take advantage of all the technological developments first achieved in animal and human genetics. Genomic analysis has demonstrated the large number of genes shared between even distantly-related species. This "synteny" means that, by working on just a few species, essential information can be found concerning a very large number of cultivated species. However, sequencing research should be focused on certain major agronomic species, either because they have a very complex genome like wheat does, or because they are further away on the phylogenetic tree from species that have already been sequenced, as is the case for beets. The costs of sequencing and marking will continue to decrease, which will improve our ability to describe the genomes of a much larger number of species and hopefully mean that the species currently considered "orphan species" will not remain so for long [25].

Statistical and mathematical methods Producing these data sets, whether for phenotyping or genotyping, requires a growing number of adapted statistical methods. To analyse morphological data, models for growth and development are used to aggregate data. For the analysis of genetic data, research in genetics and the genomic regions involved in determining quantitative traits are supported by increasingly powerful methods [26].

Marker-assisted selection and genomic selection Marker-assisted selection makes it possible to increase the occurrence of favourable alleles in the populations being selected, but exploiting all this data has led to the emergence of a new selection practice called genomic selection, the idea for which was first developed for animals in the 1990s by Lande and Thompson [27] and was truly implemented by Meuwissen and his team [28]. If we know the effect the environment has on the expression of traits, it should be possible to directly select for the genotype, which will have been identified thanks to the different markers located throughout the genome, some of which are closely linked with the genes of interest. For the same cost, we should then be able to work more quickly and/or on a larger number of characters, such as resistance to disease, as long as the genes responsible for them have already been identified. This methodology, already in practice for animals, is only now being developed for plants because of the number of plant species concerned, the complexity of plant genomes and the size of the populations being selected for. Genomic selection has already been implemented for corn and represents a real revolution in selection practices. It gives us hope that it will be possible to maintain the rate of genetic progress that was reached in the last few decades, even for species that have been the subject of much research and for which many fear no future progress is possible.

Increased Use of Irrigation?

In order to ensure the success of productive, ecological agriculture, genetic and biotechnological innovations will have to be used in conjunction with agronomic innovations. The expansion of irrigated areas is one other method that may be used in order to maintain—or indeed accelerate—the current rate of increase of production. However, in France as in most of the world, two obstacles stand in the way. Firstly, water is a limited resource, and will be even more so in the future. In many regions, it is impossible to use more water without affecting non-agricultural uses and/or the quality of aquatic environments. However, using modern techniques of water supply and optimising volumes and schedules ensure significant water savings. Thus, the amount of water used to irrigate 1 ha of maize was reduced by 30% in southwestern France, in reaction to constraints imposed on consecutive water withdrawals, especially during the 2003 drought. As INRA's collective scientific expert report of 2006 on "drought and agriculture" [22] showed using the example of the Charente region, the strategy of creating additional resources by establishing new water retention units can be used in locations where water resources are sufficiently replenished by excess water in the winter. This strategy would be even more justified, at least from a quantitative point of view, if climate change led to both an increase in winter rainfall and an increased intensity of summer droughts. The ecological impacts of this type of strategy must be rigorously taken into account and the limits strictly managed, in order to avoid returning to a level of water withdrawal that is greater than availability. Secondly, the high cost of investing in retaining materials and in water distribution must be offset by an increase in agricultural production. This requirement does not encourage savings for other inputs. Achieving better environmental performances of irrigated agriculture is not impossible, but it will depend on the amortisation of higher fixed costs.

A Multifaceted Approach

The role of French agriculture in increasing overall agricultural production while protecting the environment will depend on modifying practices and systems of production which call into question the current dominant models. A new diversification of crops and uses of rural spaces is key to efficient resource management and the supply of ecosystem services, especially those that depend on biodiversity. This diversification is, unfortunately, very often a factor of lost volumes of production and a shift toward crops that turn a smaller profit. Agronomic innovation, plant breeding and biotechnologies are necessary to reduce this tension. Once this condition is met, the agricultural supply chain—from the agri-supply industry to agribusinesses (first and second processors)—will once again be profitable, and the benefits of increased harvest volumes, market growth for new inputs and ecological benefits consistent with societal expectations reconciled. This new cohesion could contribute to lifting the "lockings"[19] caused by the currently conflicting interests of different supply chain operators. Choices made by distributors and consumers will play a determining role through the value they attribute to production methods and environmental conservation and the importance they give to product quality. The acceptance of new technologies and processes that ensure a reduced level of environmental and

[19] These lockings are the result of interdependence between these operators. For example, new production practices depend on adapted varieties, which are not available because demand from farmers and processors is not high enough for there to be a market for them (INRA 2010. Ecophyto R&D Study, Analysis of player strategies. Op. cit.).

health risks will also be decisive. The stakes are high and no opportunity for innovation must go unexplored, provided it includes an evaluation that integrates all benefits and all impacts on production, ecosystem services and on risks that may affect product quality.

It is therefore not a question of opposing a biotechnological point of view, symbolised by the development of a GMO-based agriculture, and an ecological point of view, symbolised by organic agriculture. Given the current state of affairs, neither solution would completely satisfy the needs of a productive, ecological agriculture, and neither would suffice on its own to feed the world in a sustainable fashion. The genetically modified varieties currently available to farmers have not shown, in the countries where they are used on a large scale (especially the United States), decisive advantages in terms of yields, or even efficient pesticide use. Resistance to herbicides actually results in an increased use of these inputs over the long term, especially by encouraging weeds, which are resistant to the active molecule. However, resistance to pests is indeed a factor of reducing insecticide treatments that merits study. Because organic agriculture has a lower output per hectare than so-called "conventional" methods of production, it uses more land. Organic products may thus have a worse score in terms of overall GHG emissions, because of the amount of land they consume, even though the absence of synthetic nitrogen fertilisers results in lower emission levels. A commitment to innovation thus involves a better integration of the scenarios that each of these different types of agriculture presents.

Insert 8.5. GMOs and Food Security

GMOs: a controversial technology According to the European directive 2001/18/EC, a genetically modified organism (GMO) is an organism, with the exception of human beings, in which the genetic material has been altered in a way that does not occur naturally by mating and/or natural recombination. According to this definition, organisms obtained through sexual reproduction cannot be considered GMOs. In common usage, GMOs are organisms whose genome has been altered by artificially integrating one or several DNA fragments of other species. Their development has provoked heated debate in certain parts of the world. Motivated by concern for economic risk, notably the monopolistic behaviour of the companies in question, and by considerations of inappropriate ownership of living organisms through patenting, these reactions have led the different players to take hard-line stances when it comes to sharing biological resources. This has also occurred because of ethical and sometimes ideological considerations. The most common GMOs today are plants that have been made more resistant to insects through the incorporation of a bacterial gene which produces a toxic protein, and plants resistant to an herbicidal molecule. These plants are controversial because of the possible risk of ingestion of these toxic proteins by the consumer, the risk that the modified genes could spread to wild plant species and the possibility of genetic contamination of organic plots (which forbid the use of GMOs), among other arguments. The rejection of GMOs, or at least the general mistrust of them, means there is great uncertainty about how they may contribute to the evolution of agriculture, at least in certain parts of the world, including the European Union.

Viewpoints

- Perfecting the technique

The first GMOs were developed using rudimentary technologies, which left significant traces in the genome, such as a gene resistant to an antibiotic that was used to identify gene transfers. The day will soon come when it will be possible to undertake the desired modification with "perfect" exactitude, changing one single base among the billions contained in the genome.

- Potential applications

Adding the elements of a new biosynthetic pathway to the genome of a plant could add a new function or a new trait to a variety, which could profoundly modify its use. Transgenesis has made it possible to produce Plasmodium antigens in order to develop protection against malaria in a starch-producing plant (PLoS ONE, December 2010). When this same process is applied to banana trees in order to ensure access to this type of vaccination for dispersed African populations without access to services or markets, the benefits suddenly become clear. Or take the example of rice with increased levels of provitamin A in order to better nourish the millions of people with vitamin deficiencies. Plant resistance to viruses is a key issue for GMOs, when efficient alternative methods of protection applicable in the field do not exist. Sharka, a viral disease that attacks fruit trees of the genus Prunus (peach, apricot and plum trees) in several regions of the world, merits considerable research. The study of viral genomes and resistance mechanisms has allowed scientists to design methods of producing resistant varieties using transgenesis and conventional selection techniques. A resistant transgenic line is now being studied in the United States. Sorghum's resistance to drought certainly did not appear overnight but through a series of progressive adaptations, each necessary in order for the next one to occur. This process seems to be difficult to bring together in one act of genetic transformation. However, many private and public organisations have announced that, thanks to genetic transformation, there will soon (in several years) be maize varieties available that are more resistant to drought. And IRRI, the International Rice Research Institute, is working to integrate genes in the rice genome from other grasses, which would make it more productive overall by changing its photosynthesis pathway. Questions thus arise about the distribution of benefits, environmental impact, the sustainability of the added function and agronomic support.

- The future of conventional plant breeding

Thanks to the domestication of living species, science has been able to select entirely new and extremely varied forms of organisms. The variety of dogs or horses bred from wild species is simply astounding. Likewise when it comes to plants, tropical and subtropical species gave way to the selection of very productive forms that thrive in temperate climates: need we be reminded of where wheat, maize or even sorghum came from? Conventional selection (as opposed to transgenic selection) has, for many decades now, allowed us to improve the productivity of most cereal species by 1 % per year, with more than half of this attributed to "genetic value". There is considerable diversity available in gene banks. And crossing genes allows us to generate even greater diversity and go beyond the extreme values found in these collections. It is in this way that, using the maize population available in France in the 1960s (adapted to southern France), it was possible to select types of maize which matured earlier and were more resistant to the low temperatures of maize silage crops in Sweden. Genomic analysis tools have bolstered the speed, precision and efficiency of conventional selection techniques, by identifying the genetic bases of observed variations. And there are numerous species that have not yet been analysed in this manner. Global food production relies on a minutely small number of species in comparison with the species identified all over the planet by people for different uses and properties. The successful domestication of certain species will boost our understanding of this process and reproduce it for other species more efficiently in terms of both biological as well as organisational and partnership bases.

Do GMOs have a role to play in food security? Food security will probably be made possible through the use of better adapted varieties, but also through distribution circuits that allow more open access to the technologies of ecological intensification, regulatory and political mechanisms, etc. This process was carried out very strictly in the framework of the international "biofortification" programme, which aimed to enhance the nutritional quality of diets accessible to the poorest populations. A series of analyses associating approaches from fields as varied as human health and nutrition, food technologies, production chain economics, agroecology, satellite geography, plant physiology and genetics, among others, made it possible to prioritise the constraints and actions based on the region concerned. On this basis, selection goals were identified for iron, zinc and provitamin A contents in beans, millet, wheat, maize, manioc and sweet potatoes. For almost all of these goals, genotypes carrying at least the target level of these nutrients were identified in the gene bank. Thanks to this, conventional genetic improvement was quickly able to develop varieties that surpassed these target values. Only the assimilable iron content in rice was not found in genetic resources, which meant transgenesis was used and now it is possible to produce plants rich in iron. The application of targeted research makes it possible to identify specific goals of genetic progress. The rigorous analysis of the diversity of natural genetic resources often gives access to highly valuable sources of diversity. Genetic engineering is only occasionally necessary in order to transfer new traits to plant material.

Knowledge of the issues, the expected benefits, the various beneficiaries and the biotechnical challenges at hand allows scientists to make an informed, joint decision on whether or not to produce a specific GMO to achieve this goal.

A Return to Territory Management

In order to develop an agricultural system that limits the use of environmentally unfriendly inputs and produces ecosystem services, a "rediversification" of crops and landscapes must take place. Reintroducing certain crops, such as legumes, in areas where they had been abandoned or expanding the surface area of permanent grasslands will depend on their perceived value for livestock farming, particularly herbivores. Because livestock runoff will be used for crop fertilisers, environmental benefits will be accumulated. The reassociation of plant and animal productions in the same territory is one way of improving the environmental function of agriculture. This return to the production specialisation of farms and agricultural spaces which has characterised agricultural evolution is a territorial issue. Its collective dimension is more important than the individual consideration of farms and the return to a mixed crop-livestock farming model. The scope of thinking must be broadened to go beyond the level of individual farmers. Changes over the past few decades have been accompanied by a very strong structurisation of the downstream sector, which collects and transforms animal products using large scale tools, particularly in the dairy sector, the economic efficiency of which requires widespread availability of the raw material, both in terms of distance and regular supply over time.

The collective dimension must thus be understood by farmers in the same territory, but also by farmers and players in the supply and collection industries. Involvement goes beyond the agricultural and food industries. Inhabitants of rural areas (because of the role agriculture plays in development), other activities that use land (such as forestry), consumers and companies and citizens concerned by environmental resources are all involved in different ways. The issue of quantitative and qualitative management of water resources illustrates this clearly. The role of territories, which we strive to optimise in conjunction with the productive functions of agricultural goods and the supply of ecosystem services, is a decisive one. In regions of the world favourable to agricultural production, maintaining agricultural activity already requires (for example, in mountainous regions), and will require to a higher degree in the future, the compensation, via public assistance and market mechanisms, for ecosystem services such as carbon sequestration, high-quality water supply or the maintenance of an attractive landscape. From this, we arrive at a geographical differentiation of agricultural functions. Will this differentiation between zones which encourage certain ecosystem functions and zones where the priority remains output levels be more beneficial than a strategy which aims to homogenise all services? Placing the international problem of land sharing (see Chap. 6) in a national context means working with the constraints of local environmental issues and territorial debates. It is no less legitimate on the national level than on a planetary level.

Intensive, Capital-poor Agriculture: The Case of South-East Asia

Asia, which is home to three quarters of the world's farmers and one quarter of its arable land, has shown a strong progression in agricultural production over the last few decades. Even though this continent continues to have the highest concentration of people living below the poverty line, considerable progress has been made in the fight against poverty. Agricultural production and average availability of food per capita have improved dramatically over the last 40 years, especially for rice, which is the staple of the Asian diet. Besides this extraordinary increase in food production across the continent, South-East Asia in particular has become a major exporter of several agroindustrial products (e.g. manioc, palm oil, rubber, wood, coffee, cacao).

This new situation has mainly arisen because rural farmers were able to take advantage of profound land and agrarian reforms in order to innovate, invest and increase production. Even in countries that are becoming increasingly industrialised, agriculture is still a powerful driver of growth and job creation. The clichéd view of this agriculture is that of small agricultural areas with genuinely remarkable performances, embodied by intensive, irrigated rice cultivation in rich alluvial deltas or on spectacular terraced fields. But this is not the only kind of agriculture in these areas; vast regions are also covered by rainfed crops, sometimes under limited hydric conditions or conditions of low fertility. The intensification process that has occurred in the last 40 years has affected all of the different crop systems there. In all cases, farmers have gotten the most out of small areas, by combining crops, livestock and fish production, or by diversifying their productions in order to improve their situation.

The expansion of agricultural land and agricultural intensification, responsible for increased agricultural production, are now limited by two factors. On the one hand, it has become difficult to extend agricultural land, which encroaches on fragile mountainous regions or tropical rainforest, without destroying the remaining large tropical forest areas. On the other hand, strong intensification, based on the large-scale use of fertilisers, pesticides and highly productive varieties cannot continue, because of its impact on the environment, but also because yields are plateauing and no longer ensure economic profitability. For highly productive crops, such as rice in the Mekong Delta (the increased yields of which made Vietnam a world leader in rice exports), the levelling off of yields, the cost and impact of wide-scale fertiliser use, the reduction of agricultural land due to urbanisation and, in the long term, rising sea levels all mean that farmers will have to ensure the sustainability of their crop systems by maximising the efficiency of inputs, including water, the use of which is becoming increasingly competitive.

In this context of highly "technified" and reactive farmers, research takes on a vital role. Despite our ability to create ever more productive varieties capable of breaking current records and to better adapt these on a regional level, we must work toward a form of ecological intensification that will allow for more efficient use of fertilisers, supervised control of pests, or even more efficient use of labour because wage gaps and urban migration have paradoxically led experts to worry about a lack of farm labour in the future. Most of the terms evoked to describe French agriculture are applicable here, particularly the "reconception" of crop systems, the integration of ecological services and the systematic analysis on the landscape level, but in a context of low capitalisation in which labour is still abundant, despite a strong movement towards urban jobs. For rainfed agricultures, which are still far from achieving their potentials, advances in yields are generally linked to their ability to optimise the management of hydric resources, to improve fertility and to optimise specific diversifications. The basic principles of ecological intensification (increasing biomass, making systems more complex and optimising geochemical cycles) must be the founding principles for improving systems, especially given that the financial capacity of farmers for buying inputs is limited. Particular efforts in selection and diversification should be made in order to serve these principles. Integrating agriculture and livestock farming or agriculture and fish farming is quite common in these regions, which in some cases have developed high-performing circular systems of raw materials.

Agricultural Systems with Productivity Reserves

The following pages will analyse agricultural systems which have access to the resources of the first green revolution and can use these resources more efficiently to greatly increase production. This is the case, for instance, of Russia, Ukraine and Kazakhstan. Wheat production in Ukraine, which currently stands at around 3 t/ha, is far from what it could be, considering the country's favourable soil and climate conditions. This performance deficit is mainly due to local conditions of agricultural development, including the poor economic organisation of farms and supply chains, and lack of infrastructure and training of players. Despite the very real difficulties in overcoming these obstacles and the necessary investment involved, we can reasonably expect to see at least some improvement in the future. Because these countries are important exporters of field crops, the level of progress will depend on prices in world markets. The current situation of abundant land (in absolute terms and in terms of agricultural labourer) translates economically into lower goals for yields than those in Western Europe, and may contribute to limiting the use of costly chemical inputs. However, this does not allay fears about the environmental effects of the changes expected to occur. Considering the vast areas concerned, it is obviously preferable that agricultural production be increased in an ecological manner. While the most extreme examples, former Soviet countries are not the only ones with reserves of land productivity. In countries with large farms, especially ones that are up against competition in world markets without public assistance, management of production costs leads to a more efficient use of inputs, particularly fertilisers. Given their expenses, these agricultural systems can only be considered partially intensive. This efficient optimisation directly depends on the relationship between the price of agricultural products and the price of inputs. Once again, it is the price variation that will determine the possible growth of yields in South

America's exporting countries (Argentina and Brazil) as well as for certain crops in highly-productive countries such as the United States.

Is Sub-Saharan Agriculture in Need of New Development?

The agricultural systems of developing countries have greatly evolved over the last 40 years. Most of these countries were able to take advantage of the green revolution, which associates productive varieties and high levels of inputs, with the successes and limitations that have already been analysed here. In sub-Saharan Africa, the green revolution did not significantly affect the productivity of agriculture. Even though the capacity of farmers in this region to adapt and innovate has been proven by significant growth in average yields (see Insert 8.1), given the particularly difficult context of the disorganisation of public assistance, yields more often remain below their potential, especially for subsistence crops. The result is that, given the continually strong demographic progression, the evolution of agriculture has not fundamentally changed the situation in terms of poverty and food supply. In these regions, there is a considerable potential for productivity gains which would moderate the need to expand arable land in order to increase production. It is clear that the potential of cultivatable land is significant, but when we consider the issues of access, conservation and conflicts of use, it is preferable to intensify the currently cultivated land rather than expand, given the negative externalities that this would provoke. Managing the conditions necessary for realising the potential of these lands will not rely on new technologies to be invented, which was the case for Asia's green revolution. Guaranteeing a stable land base, storage infrastructure and access to markets, inputs and credits, as well as technical support are decisive factors for the success of intensification throughout the world. The lessons to be learned from the green revolution—whether in terms of environmental impact, erosion of the productive potential of ecosystems or more broadly of agrarian systems—will force us to explore new methods of intensification.

We must try to learn from all of the lessons—both positive and negative—of the first green revolution, especially where the following two issues are concerned: increasing yields while limiting negative environmental impacts, and doing so while making the most efficient use possible of inputs—whether they are fertilisers, pesticides or energy—which must be used parsimoniously because of their cost and finiteness. It is because of these two constraints that we must strive to find ecological ways of intensifying agriculture, but they are also what make it possible to imagine achieving a "doubly green" agriculture without first going through the steps of the "singly" green revolution. In addition, African demographics make it necessary to consider maintaining and improving on agricultural jobs as a necessary objective of intensification. We cannot develop a form of modernisation where local rural populations are evicted, under the pretext of economic viability, because we know that in most cases evictions would increase the number of urban poor and deal a blow to food security. It is therefore highly preferable, for agricultural reasons, but also for social and political stability, that agricultural intensification be carried out in a way to allow the value of labour to be recognised. In such an intensification process, the role played by technology will, of course, be important, but the general economic context, the ability of leaders to develop and oversee policies favourable to agriculture, the organisation of markets and the different regulatory tools will be equally significant and decisive factors for success, and will be employed for the long term and allow for a true mobilisation of farmers. It is clear that farmers will have no reason to increase output if their access to the market is erratic and unprofitable; likewise, they will not want to invest in the fertility of their soil if there is no long-term guarantee for their land. In addition, the financial capacity of most farmers means they will not be able to use expensive technologies or inputs, which is another reason to choose ecological intensification.

Finally, since we will be imagining "custom-made" solutions rather than "ready-made" ones, the role of farmers in innovation will be essential and must be assisted by continuous research, training and technical support. Farmers will be the ultimate "engineers" of technical know-how and the precise knowledge of a given environment and its resources, and will strike the right balance between technology and technical decisions. This "knowledge-intensive" form of agriculture clearly requires shared information and the training of farmers, as well as a more participative stance on the part of researchers. In developing countries, the instability of the climatic, economic and regulatory context of agricultural production and the scant resources in family farming, mean that production and transformation systems must be highly adaptable, which in turn requires broadening the range of production conditions to be envisioned. In order to imagine the different technical drivers of this new intensification, we must try to better understand its deciding factors. If we limit ourselves to plant production, yields (or productivity per land unit) can be considered the result of the photosynthetic production of a plant cover during a given period, using resources from the surrounding environment or plot. We can then see that what is at play is the interaction between a plant cover, with one or more genotypes, an environment characterised by a certain climate and available resources and a certain method of cultivation.

In order to optimise yields, it is thus necessary to work on each of these elements of the equation, as well as—and especially—their interactions, because these have a particular importance in the intensification process. It would therefore be counterproductive to use large quantities of fertilisers for plants that have not been selected for this or to plant elite varieties too sparsely or in nitrogen-poor environments. If intensification is carried out properly, the logical progression would be to first improve the crop and environment management before trying to achieve the full potential of the improved varieties. This is why it is unrealistic to seek to increase yields by using improved species when it is not certain that farmers will be able to use the appropriate fertilisers, which is clearly the most limiting factor of yields in sub-Saharan Africa. There are no "miracle" varieties. However, there are numerous agricultural success stories in this region, based on innovative technologies used in conjunction with improved varieties and technical assistance, which proves that farmers respond to new ideas that benefit them economically. Export crops such as cotton, coffee, oil palm and rubber tree usually benefit from considerable scientific, financial and institutional investment, for which integrating the supply chain has made it possible to fill in the gaps in infrastructure and organised markets. Traditional subsistence crops (millet, fonio, cowpea, yam, manioc, etc.) as well as more recently-introduced crops (rice, maize, wheat, etc.) have not benefitted from such significant investment, especially since the political decision-makers and development donors have long thought it was preferable to import these staples from developed countries. The basic principles of ecological intensification in developing countries will involve working with all the different drivers for action—be they agri-technical, organisational or genetic—which must be used conjointly in order to create innovative systems.

Thinking and Acting Globally

In order to examine agriculture's dual contribution to feeding humanity and preserving the normal functioning of the biosphere (which in turn favours agriculture), we must look at the issue from a global point of view. This means not only taking into account the different levels of territorial organisation, from neighbourhoods to the planet, but also all the functions and consequences of production systems. Agricultural production has now outgrown the paradigm of the green revolution, which consisted in optimising the productive performance of industrial inputs, making the environment as artificial as necessary and minimising losses by eliminating pests and diseases. Glasshouse crops, which use an artificial growing medium and nutrients in solution, are one illustration of this. For example, we know how to greatly reduce plant diseases by managing the internal climate and using crop practices and biological control. This reference to manipulating ecosystems should not be taken as a comparison to streamlined, output-oriented factories. However, there are similarities, such as the focus on performance and some of the improvement principles. What differentiate the two are integration and the systemic vision. We can no longer separate this agro-ecological approach from the economic, social, organisational and changing political conditions that will allow us to develop it. It is necessary to place things in a context of sustainable development, which must make sense from the most basic to the planetary level. Due to a lack of regulatory mechanisms, the primacy of territorial interests in decision-making does not work to optimise the management of global public goods. Without ignoring the substantial action in favour of developing these regulatory mechanisms through international conventions, local action must be conceived of in relation to its effects of these shared goods of humanity[20], i.e. the atmosphere, biodiversity and soil. Not only must we think

[20] Even if the status of public goods is not protected by universal laws, this is a very real dimension, insofar as the ecosystem services they provide are planetary or concern a planetary market. Water does not fall neatly into this category, because its functions are territorial, but in this context it has a similar status because of the role it plays in food production. This is what supports the concept of "virtual water" which refers to water used for irrigation and water consumed by rainfed crops, and includes the "hazardous" consumption by all plant cover, whether or not this is consumed by people or livestock.

globally, but local actions must be oriented towards a positive contribution to the global future.

Committed Innovation to Fulfil Needs

The food challenges and ecological challenges the world faces are severe. They can only be met by successful innovation which requires significant commitment. While research is necessary, innovation is not solely dependent upon it. The model of the green revolution—that is, technical innovation which was then deployed at the local level in simplified terms for the user—has been reformatted. It will now be necessary to organise the processes of the functioning of "innovative networks", which will be interactive and involve research, structures for technical and agricultural support, companies, farmers and contributors. In France, this has already occurred: there are now groups that bring these players together to design innovative crop systems and outline their development. INRA's Ecophyto R&D study showed that it is more efficient to combine different strategies of phytosanitary protection than to move towards a uniform strategy. We must pursue the benefits provided by the evolution of agricultural practices, by animal and plant breeding, by biotechnologies, by better controls of ecological processes or by the creation of bioproducts with low environmental risks. Innovation will involve different levels of organisation. Plots, which remain the main focus of agronomic studies, will be integrated in a territorial context or a functional area for management of water or biodiversity. This should extend to the planetary level. In the same way, progress in the functioning of farming should be extended to progress in the supply chain of which it is a part (agri-supply and product distribution) in order to imagine viable innovations. This necessary vertical integration will be added to the necessary research in horizontal—or spatial—integration. This is true for productive functions as well as for environmental functions.

Distributing the investments necessary for such an innovative endeavour is obviously problematic. The volume and efficiency of these investments requires meeting two conditions: first, an enforced coordination of international agronomic research and second, the capacity of agronomic research to include other scientific communities in the issues it tackles. These may include ecology—traditionally more focused on natural ecosystems—but also climate sciences, as well as social, management, legal and geographical sciences. Interdisciplinary scientific expertise, such as that of the IPCC on an international level, also contributes to a constructive meeting of minds. Bringing together researchers from different disciplines, which is already the case in France and throughout the world, should reinforce this dynamic, as well as the alliances between specialised scientific institutions. This enlarged, better structured mobilisation could help to identify more clearly the outlook for the most efficient and fast-acting impacts of innovations to develop. For example, it would help us to differentiate between situations where agronomic progress depends on the implementation of available knowledge and manageable technology (at the expense of local adaptation that can be carried out in the short or medium term), and those that require basic research and significant technological investment and have uncertain, long-term effects. This mobilisation must contribute to the construction of innovative networks which associate research, training, agricultural development structures, companies, farmers themselves and public managers, with each one playing a part in the systemic process of transformation to be carried out.

We now know that we must move beyond the linear "chain" of innovation, in which public and private researchers imagine the innovation that they will then need only to distribute. It is no longer possible to simply distribute a uniform "technological package". The issue at hand is now to precisely adapt production modes to ecological, territorial and economic contexts in all their diversity. The interactions between those who are in charge of the evolution of these processes will be of the utmost importance. Specialised groupings by type of production (e.g. vegetables, field crops, livestock) and the creation of farm networks to implement the Ecophyto 2018 plan are moving in this direction. They must include the other societal players besides those concerned by the transformation of agriculture. Furthermore, the need for innovation also implies evaluation methods and the acquisition and management of the information necessary for their implementation. Initial evaluations using forecasting tools are being used more and more to support public decisions and international negotiations. But used in a downstream role, they also allow for monitoring and correcting the effects of the decisions made. Given the inertia of the systems to be managed (the global climate being the most well-known example), any insufficiencies may lead to consequences that are slow to reverse or even irreversible on a human scale. Adopting international standards after negotiations will potentially amplify the risk associated with these mistakes. These standards put the focus on monitoring indicators, which are necessarily reductive, and can mask real, often subtle changes.

An Unstable Context

The consequences of these changing conditions will depend on geopolitical, economic and environmental factors. A similar, obvious remark can also be made for countries, whose history of development has not allowed for productive agriculture, by implementing production systems which ensured the success of the green revolution. The level of margins that farmers can achieve will also play a decisive role. In the future, price level, more and more systematically determined in all areas by conditions of supply and demand in increasingly connected, globalised markets, will play a decisive role on a planetary level. Price instability poses obvious problems for long-term investments, but also for a dynamic of continual change for which the importance of food and environmental goals could be greatly affected.

A long-term hike in agricultural prices would encourage progress in agricultural systems that remain far from their potential, and allow for development in small farms currently in economic survival mode and among the first victims of food deficits. This would also encourage intensification strategies favouring the use of inputs that are instantly efficient and profitable, to the detriment of the environment. It would increase the difficulty of access to food for malnourished, non-agricultural populations, at the very least in the absence of sufficient economic development to increase wealth (see Chap. 9). The effects of price levels and their variations on the new agricultural revolution to be carried out are contrasted. As a factor of instability in the markets, extreme environmental events allow for increased awareness and a change in behaviour on the part of many players, from farmers to public officials. In France, the 2003 heat wave and drought helped raise awareness of the risks of climate change. The agricultural and food crises of 2007–2008, partly provoked by climate factors, led to investment projects—or land grabs—in land and agriculture in countries where it was available (see Chap. 7). The occurrence of droughts and devastating hydraulic or erosive events, as well as the need to make up for the absence of fruit pollinators, will affirm the value of the ecosystem services provided by agriculture (see Chap. 6). The strategic issue associated with the capacity to anticipate extreme events and their consequences is, from this point of view, decisive. Finally, international agreements, and their national implications, regulatory tools and policies for implementation are obviously major events favourable to change. The uncertainty of their continuity and their low consistency contribute to the wait-and-see policy, increasing the dangerous inertia and the contradictions between local strategies and global necessities. Agriculture's contribution to sustainable development will depend on a more legitimate and decisive global governance. International research, necessarily oriented towards sharing the universal objective of the accumulation of knowledge, must attempt to be the precursor, if not the catalyser. The example of the IPCC can serve as a guide in this regard. We will return to this point in Chap. 10.

References

1. Dumont R, Rosier B (1966) Nous allons à la famine. Éditions du Seuil, Paris, 268 p
2. Robert et Cheverry dans Stengel P, Bruckler L, Balesdent J (eds) (2009, Janvier) Le sol. Éditions Quae, 184 p
3. The United Nations World Water Development Report 2 (2006) Water: a shared responsibility. UNESCO, Paris, and Berghahn Books, New York. http://www.unesco.org/water/wwap/wwdr/wwdr2/
4. GIEC (2007) Bilan 2007 des changements climatiques. Contribution des Groupes de travail I, II et III au quatrième Rapport d'évaluation du Groupe d'experts intergouvernemental sur l'évolution du climat [Main editorial team, Pachauri RK et Reisinger A (dir.)]. GIEC, Genève, 103 p. http://www.ipcc.ch/pdf/assessment-report/ar4/syr/ar4_syr_fr.pdf
5. http://www.fao.org/docrep/010/a0701e/a0701e00.HTM
6. EEA (2010) The European environment—state and outlook 2010: synthesis. European Environment Agency, Copenhagen. http://www.eea.europa.eu/soer/synthesis
7. http://www.fertilizer.org/ifa/HomePage/MEDIA/Press-releases-2009/28-July-2009; et http://www.fertilizer.org/ifa/…/1/…/2009_ifa_climate_change.pdf
8. Seguin B (2010) Coup de chaud sur l'agriculture. Delachaux et Niestlé, Paris, 205 p

9. http://www.unep.org/Documents.Multilingual/Default.asp?DocumentID=562&ArticleID=6084&l=fr et http://www.unep.org/pdf/foodcrisis_lores.pdf
10. INRA, Brisson N, Levrault F (coordinateurs) (2010) Livre vert du projet Climator. Changement climatique, agriculture et forêt en France: simulations d'impacts sur les principales espèces. ANR-Inra, Éditions Ademe, 336 p. http://www2.ademe.fr/servlet/getDoc?cid=96&m=3&id=70992&p1=30&ref=12441
11. Jouzel J, Lorius C, Raynaud D (2008) Planète blanche. Éditions Odile Jacob, Paris, 305 p
12. Arvalis (2009) Impacts environnementaux des techniques culturales sans labour en France. Paris. http://www.editions-arvalis.fr/ouvrage.php?id_ouvrage=929
13. INRA (2010) Ecophyto RD: quelles voies pour réduire l'usage des pesticides? http://www.inra.fr/l_institut/etudes/ecophyto_r_d
14. Jacquet F, Butault JP, Guichard L (2011) An economic analysis of the possibility of reducing pesticides in French field crops. Ecol Econ 70(9):1638–1648
15. Guyomard H (2009) Politiques publiques et agriculture biologique. Innov Agron 4:499–511
16. http://www.strategie.gouv.fr/article.php3?id_article=810
17. Doré T, Maraux F (2010) Les manières de produire en agriculture, état des lieux et controverses. In: Doré T, Réchauchère O (eds), La question agricole mondiale: enjeux économiques, sociaux et environnementaux. La Documentation française, Paris, pp 115–134
18. Meynard JM (2009) Conclusion: quelles priorités pour la RD en agriculture biologique? Innov Agron 4:495–498
19. http://www.fao.org/organicag/ofs/docs_fr.htm; et http://www.google.fr/search?q=Conf%C3%A9rence+internationale+sur+l%E2%80%99agriculture+biologique+et+la+s%C3%A9curit%C3%A9+alimentaire%2C+Rome%2C+3-5+mai+2007%2C+FAO&ie=utf-8&oe=utf-8&aq=t&rls=org.mozilla:fr:official&client=firefox-a
20. Petit S, Thenail C, Chauvel B, Le Coeur D, Baudry J (2008) Les apports de l'écologie du paysage pour comprendre la dynamique de la flore adventice. Innov Agron 3:49–60
21. Simon S, Sauphanor B, Defrance H, Lauri PE (2009) Manipulations des habitats du verger biologique et de son environnement pour le contrôle des bio-agresseurs. Des éléments pour la modulation des relations arbre-ravageurs-auxiliaires. Innov Agron 4:125–134
22. http://www.inra.fr/l_institut/expertise/expertises_realisees/reduire_la_vulnerabilite_de_l_agriculture_a_un_risque_accru_de_manque_d_eau
23. Vankadavath RN, Hussain AJ, Bodanapu R et al (2009) Computer aided data acquisition tool for high-throughput phenotyping of plant populations. Plant Methods 5(18)
24. Yazdanbakhsh N, Fisahn J (2009) High throughput phenotyping of root growth dynamics, lateral root formation, root architecture and root hair development enabled by PlaRoM. Funct Plant Biol 36(10–11):938–946
25. Varshney RK, Penmetsa RV, Dutta S et al (2010) Pigeonpea genomics initiative (PGI): an international effort to improve crop productivity of pigeonpea (*Cajanus cajan L.*). Mol Breeding 26(3):393–408
26. Guo BH, Sleper DA, Beavis WD (2010) Nested association mapping for identification of functional markers. Genetics 186(1):373–383
27. Lande R, Thompson R (1990) Efficiency of marker-assisted selection in the improvement of quantitative traits. Genetics 124(3):743–756
28. Meuwissen THE, Hayes BJ, Goddard ME (2001) Prediction of total genetic value using genome-wide dense marker maps. Genetics 157(4):1819–1829

Chapter 9
Feeding the World Starts with Fighting Poverty

The riots that erupted in several world capitals when international food prices surged in 2007–2008 brought the issue of hunger to the fore once again, making it a top priority for the international community. Shortly after these riots, numerous declarations and propositions insisted on the need to bolster agricultural production. However, these food riots and the public debate it spurred only provide a partial view of the problem of hunger and food security. This chapter aims to provide a more detailed analysis of hunger, by outlining the issue and providing an overview of its causes and related factors. As with other chapters, it closes with possible courses of action to chart.

Mapping Hunger

In its 2010 report on global food security, the FAO [1] estimated that 925 million people in the world suffer from hunger. This figure represented a decrease of 100 million from 2009 (1,025 million), after rising for two consecutive years (100 million per year), in 2008 due to higher agricultural prices and in 2009 due to the banking, financial and economic crisis. Before analysing what has caused this situation, which is universally recognised as unacceptable, let us try to better understand what lies behind these figures. What is the reality of hunger today and how can it be described and measured?

Hunger can take on several forms: acute hunger, which occurs when it becomes impossible to access food for different reasons (e.g. extreme weather events, job loss, war); chronic hunger; qualitative nutritional imbalances; nutritional deficiencies; etc. People are generally considered to be in a situation of food insecurity when they do not have physical, economic or social access to sufficient quantities of healthy food to meet their energy and nutritional requirements. The FAO's estimates are calculated using purely calorie-related data: after calculating the theoretical in-

M. Guillou, G. Matheron, *The World's Challenge*,
DOI 10.1007/978-94-017-8569-3_9, © Éditions Quæ, 2014

dividual availability[1], the probable number of people living on less than 1,800 kcal per day (which is considered a situation of food insecurity) is extrapolated. This figure, calculated on a national basis, is based on several assumptions about distribution which are difficult to verify because consumption surveys are seldom precise. Even though it is clear that these figures must be used judiciously because they do not represent precise measurements, they remain extremely useful in gauging different situations and are taken into consideration when drawing up food policies.

Other, more qualitative indicators exist which stretch the scope of the hunger issue. For example, the Global Hunger Index (GHI) combines three indicators as percentages: the proportion of undernourished people (FAO data), the proportion of underweight children below the age of five (WHO data) and the child mortality rate (UNICEF data). The GHI has the advantage of taking into account other important aspects of hunger besides calorie availability. Indeed, food security issues are not limited to calorie intake. For example, the statistics for people suffering from micronutrient deficiencies are alarming: in poor African and Asian populations, on average less than 50 % of people can satisfy their physiological needs in terms of micronutrients (WHO). From a public health point of view, the most consequential deficiencies are usually iodine deficiencies, a major cause of delayed brain development (740 million people); vitamin A deficiencies, which are responsible for immune disorders and impaired vision (140 million children under the age of five); and iron deficiencies, which can lead to anaemia and delayed psychomotor development (2 billion people, mostly women and young children). In terms of absolute numbers, the situation in South-East Asia is particularly worrisome.

The different hunger indicators generally show the same trends. The 925 million hungry people in 2010 represent a statistical image of hunger, calculated from an overall, indirect count. However, it is particularly difficult to imagine the gravity of the situation simply by citing an increase or a decrease of 100 million hungry people. But if we accept these statistics as indicators, we can outline more clearly the current "geography" of hunger. First, however, it is worth reiterating that malnutrition can stem from non-food-related causes. There are many contributing factors. Child malnutrition, which delays growth and development with often irreversible consequences, is as much a result of insufficient food as insufficient medical care: diarrhoea and malaria are the leading causes of delayed growth. The "Sikasso paradox" is a clear example of situations where malnutrition increases even though per capita food production increases. The reasons behind this are twofold: overall growth in production and revenues from export crops is not enough to pull large swathes of a population out of poverty [2], and women who work longer in the fields are less able to give their children the care and nourishment they need. The geography of hunger is thus certainly tied to food availability and a population's access to food, as well as hygiene, health, maternal care, education and generally all the factors linked to poverty [3].

[1] Theoretical individual availability is calculated by adding production and imports, then subtracting exports and losses, before being compared to the calorie requirements of the given population, taking into account its structure.

Giving a Face to the Hungry

Poverty rates are highest among rural populations [4]. Nearly 80 % of those suffering from hunger are rural poor, with 50 % accounting for poor peasants, 22 % landless peasants and 8 % people living in communities with traditional lifestyles, such as nomadic herders, family fishermen or forest dwellers. The remaining 20 % is comprised of urban poor, whose numbers are rapidly increasing [5]. Beyond these figures and averages per country, who are those suffering from malnutrition? National surveys show that women and children are especially affected. Depending on the location and time of year, 10–40 % of women have a body mass index (BMI) of less than 18.5[2], which is a sign that they are underweight. This can provoke a vicious circle, by causing low birth weights, which in turn lead to a higher incidence and prevalence rates of disease and mortality in young children, and delayed growth from an early age, which is at the root of motor or cognitive developmental problems at a young age and throughout life.

Putting Hunger on the Map

Statistics calculated by country allow us to pinpoint malnourished populations. Among the 925 million people identified as hungry in 2010, it appears that 578 million are in Asia and the Pacific (15 % of this region's population), 239 million in sub-Saharan Africa (30 %), 53 million in Latin America and the Caribbean (9 %), 37 million in the Near East and North Africa (8 %) and 19 million in industrialised countries. Seven countries (Bangladesh, China, the Democratic Republic of Congo, Ethiopia, India, Indonesia and Pakistan) are home to two thirds of the world's hungry [1]. Half of all low birth weights in the world are found in South Asia (particularly India and Bangladesh) where the prevalence is 20–40 %, and in sub-Saharan Africa (around 15 % on average). The prevalence of underweight children aged 0–5, which is highly correlated to the prevalence of low birth weight, is around 25 % in Asia, 22 % in Africa and 4 % in Latin America. In terms of absolute numbers, simply because of its population, Asia's situation appears the most troubling, but it is in Africa that the problem seems to have taken strongest hold. Considerable progress has been made in some countries in terms of nutrition, and especially childhood nutrition. According to the Global Hunger Index, between 1990 and 2010, South Asia reduced its hunger index by 25 % and sub-Saharan Africa by 14 % [6]. Several countries have thus achieved the first of the Millennium Development Goals, which consists in halving the number of undernourished people between 1990 and 2015: in Africa, these countries are Congo, Ghana, Mali and Nigeria; in Asia they are Armenia, Myanmar and Vietnam; in Latin America and the Caribbean they are Guyana, Jamaica and Nicaragua. Countries as varied as China, Ethiopia and Brazil

---

[2] Remember that BMI is defined as a person's body weight (measured in kilograms) divided by the square of his or her height (in meters).

are very close to achieving this as well. However, in other countries, the situation has deteriorated, sometimes because of war, but also for other reasons, such as corruption and poor economic choices and policies. For example, in the Democratic Republic of Congo, two-thirds of the population are currently considered malnourished.

As we have already seen in Chap. 2, developing countries are faced with a double challenge. In the same country, the seemingly incompatible issues of overnourishment and undernourishment exist side by side, and, just as in OECD countries, there is also an increased occurrence of diseases linked to excessive nutrition, in particular an unbalanced diet and lack of exercise. Once again, the causes are complex and the so-called "epidemic" is not solely the result of diet or food transitions. Taking action in this context of disparity remains a highly complex matter: subsidising high-calorie foods in order to give the poorest populations access to them could provoke overconsumption by other poor populations [7]. In industrialised countries and countries with moderate incomes, malnourishment due to insufficient calorie intake and/or deficiencies is on the rise as the number of poor households increases. In France, more than 1 million people report being hungry and 2.5 million receive food aid. These figures are comparable to those of poverty, with 4.3 million people when the poverty line is set at 50 % and nearly 8 million when it is set at 60 %[3]. The ranks of the poor have swelled since the beginning of the century after decreasing throughout the last three decades of the twentieth century[4]. Among other examples, 17 million people in the United States are in danger of finding themselves below the poverty line.

The Usual Suspect: Blaming Malnutrition on a Lack of Available Food

The lack of food, whether locally-grown or imported, available to consumers is sometimes responsible for temporary food crises, which may be caused for example by extreme weather (e.g. drought, flooding) or a sanitary crisis (e.g. plant pest, animal epidemic, locust invasion). These situations are also, unfortunately, very common during times of conflict. In the African Great Lakes region for example, war and food shortages take a devastating toll on populations. These crises may be global, political, economic, humanitarian, etc. Poorly organised food markets hinder the performance of the supply chain, leading to significant losses and greatly affecting entire populations—especially those who are the most vulnerable, more or less for the long term. At the same time, security measures such as international humanitarian aid, including food aid, as commendable and necessary as they may be,

---

[3] A person is considered to be poor when he or she lives below the poverty line. In France, the poverty line is usually established at 50 % of the median standard of living, while Eurostat sets the line at 60 %, which is now the more frequently published figure.

[4] The Inequality Observatory (*Observatoire des Inégalités*), consulted online on 6 February 2011.

set in motion a vicious circle by fostering dependency, slowing the reconstruction or reinforcement of local institutions and hampering the ability of local agricultural systems to rise to the issue.

Food crises may also be chronic. The FAO described the situations of 22 countries experiencing prolonged crises for various reasons: long-term wars, prolonged failure of the government and institutions, successive or large-scale natural disasters, etc. These 22 countries have very high levels of food insecurity, which is the source of devastating infant mortality rates, delayed growth of a large portion of the population, emaciation, etc. The most shocking example is undoubtedly the Democratic Republic of Congo, where the percentage of undernourished people went from 26 % in 1990–1992 to 69 % in 2009–2010. In the 22 countries studied, the proportion of underfed people stands at 37 %, versus only 13 % on average for all the Least Developed Countries (LDC) of the planet. Conflicts exacerbate the processes of structural fragilisation and impoverishment that have taken root over the past few decades and which constantly affect the poorest populations and the landless. As we have already seen, the latter constitute the largest proportion of hungry people, because of low yields, post harvest losses, vulnerability to risks, low revenues generated by agriculture, the impossibility to generate resources during lean periods, etc.

The Link Between Food and Poverty: Common Assumptions and Misconceptions

The acute or chronic lack of available food, due to production or supply deficits, often makes headlines. As dramatic as the situation may be, the world often forgets that, in the vast majority of cases, there is enough food available to feed everyone, at least in theory. Food insecurity is first and foremost the inability for some people to access food because they simply cannot afford it. A lack of money is the most limiting factor to consumers' access to sufficient, high-quality food, and thus the main cause of systemic malnutrition. This is tied to the issue of wealth and revenue, and their geographic distribution, between as well as within countries. It is also tied to the question of economic and social development throughout the world. We must recognise the astoundingly uneven distribution of wealth in the world. On a global scale, wealth is distributed along a curve: according to the United Nations World Institute for Development Economic Research in Helsinki, 1 % of the richest people hold 40 % of global wealth and 10 % hold 85 %; at the bottom of the curve, the "lower" half of the global population barely holds 1 % of global riches. What is true at the global level is also true within most countries, whether they are industrialised, emerging or less developed. Income in France is distributed very unequally, with the poorest 10 % receiving only 3.7 % of income and the richest 10 % receiving 24.1 %, or 6.5 times as much. This inequality is certainly less severe than in many developing countries, but it still shows that even after the redistribution of wealth through taxes and social services, inequality persists[5]. The food and agricultural price crisis of 2007–2008 and the financial and banking crisis of 2009, which quickly took on the proportions of an economic crisis, did not help matters. The recent spike in agricultural prices at the end of 2010 and the beginning of 2011 is likely to make things worse if it continues and nothing is done to stop it.

The international community's response to the 2008 food riots focused on the need to boost agricultural production. This was notably the case during the FAO Summit in June 2008, the final declaration of which reflected this position: in order to respond to the crisis, it would suffice to increase agricultural production. The text insisted on the need, in the short term, for "immediate support for agricultural production and trade" and, in the long term, for investment in food production. Throughout the entire declaration, the word "production" (usually following "agricultural" or "food") occurs nine times, whereas "food access" only occurs once, as does the word "poor" [8]. Even though, as we have seen, the actual production of food commodities can be the cause of crises in certain circumstances and in certain places, it is often not the most urgent of problems, at least on a short-term, global scale (because, as illustrated in Chap. 8, it will be necessary in the medium term to increase agricultural production in order to feed 9 billion people by 2050, using a necessarily productive and ecological form of agriculture). Concretely, these food riots had as much to do with poverty as with food. The protestors' ire stemmed as much from increased oil prices (in Cameroon and Iran), inadequate salaries (Egypt) and rampant corruption and greed among the elite. Urban poverty is a common backdrop to riots, and instability or political tension (as in Guinea and Haiti), or inter-community war or conflict (Sudan, Chad, Somalia, Kenya and Pakistan) are often its handmaidens. Such riots are sometimes simply a new episode in an ongoing series of protests or violence (Nigeria, Zimbabwe, Myanmar, India, Bangladesh and Malaysia). More recently, the tragic events (in terms of number of victims) currently shaking the Arab world may also, at least partly, be evaluated in these terms. Even in Dakar, a city fuelled by the international market, where protestors wielded bread to demand fairer prices, consumer organisations blame energy prices as much as basic food prices. Food insecurity is thus first and foremost a problem of access and income, or, more precisely, the relationship between these two and prices. The very concept of food insecurity has changed over time and today highlights this basic determining factor.

In 1974, in the aftermath of the oil crisis, with skyrocketing prices and in the wake of the devastation caused by droughts and floods, the definition of food insecurity focused on food availability. Thirty years of research and debate, particularly influenced by the work of Amartya Sen who was awarded the 1998 Nobel Prize in Economics—and particularly his work on famine and economic theories of development [9, 10]—helped to complete the initial definition by adding three key factors: access, regularity and quality. Even if production remained paramount, the 1996 definition, confirmed in 2002, shifted the focus onto other drivers for action. Supplying food in sufficient quantities of course remains a necessity, but this

---

[5] The Inequality Observatory (*Observatoire des Inégalités*), consulted online on 6 February 2011.

is no longer enough. It is also necessary to ensure that each individual has access to food, through his or her own means of production and/or through the market. The fight against poverty and inequality thus becomes a means of guaranteeing food security, as the wording of the Millennium Development Goals at the 2002 Johannesburg Summit [11, 12] clearly states. Even though it concerns less people in terms of absolute numbers, it is shocking to see the growing pervasiveness of urban food insecurity. This is what causes uprisings and riots and generates political instability, as happened at the beginning of 2011 when the prices of raw commodities skyrocketed once again. Urban food insecurity does not spare the great cities of industrialised nations—far from it: in France, where homeless people often take shelter in the underground transportation system or even more precarious locations, the "*Restos du Cœur*" ("Restaurants of the Heart") charity was created and serves an ever-increasing number of meals year in and year out.

In rural areas, home to the greatest number of malnourished people, agriculture plays an essential role, going far beyond its primary function of providing access to food [13]. It also has an economic aspect that is key to development, generating jobs and income[6]. This is why, as proposed in CIRAD's initiative within the framework of the Observatory for World Agricultures (*Observatoire des agricultures du monde*—OAM), hunger must be evaluated in a broader scope, going beyond productivity, yields per hectare or litre of milk per cow, even if these criteria remain crucially important (see Chaps. 6, and 8). For rural populations, agriculture is a source of income which serves not only to buy food, but also to obtain other goods and services and to make investments. Investment in other activities generates additional income which can in turn be used to buy food. One of today's key challenges is for agriculture to continue to enable this transformation that is at the heart of the historic process of the evolution of societies. Indeed, agriculture shapes the very structure of society, leaving a lasting impact on its economic and demographic transitions, which differ according to the role societies play in globalisation. In turn, society has shaped agricultural issues and agriculture's contribution to economic and social development. The problems of rural poverty, and its local and far-reaching consequences in urban areas, are also linked to the transformation process of the agricultural sector.

The link between food security and poverty having been clearly established, common perceptions about poverty must now be reassessed and the links between poverty and development established. Going a step further, we must observe how these links in turn affect food security, and finally how agriculture fits into the picture. It is worth noting the shift in focus of the current political discourse and positions supported by international organisations, which are moving away from inequality and toward the issue of poverty and the related issues of initial investment and capabilities[7]. The Millennium Development Goals are a testament to this,

[6] Not to mention, of course, environmental and biodiversity services and functions (see Chap. 6).

[7] To use Sen's term. In practice, his theory of capabilities is used to distinguish between the concept of "functionings" and "capabilities". "The former represents what an individual can accomplish given the means at his disposal (sufficient food, freedom of movement, reading and writing)—and

as they aim to reduce poverty and go as far as developing indicators to represent this goal. This precludes a rethinking of the controversial hypothesis that holds that inequality is a necessary ingredient of economic growth. However, research in development economics currently focuses as much on reducing inequalities as on growth and development. Is it possible that there are threshold and/or ripple effects that will force us to reassess the impact of crisis and conflict and the inequalities these engender? Is a dichotomy forming that condemns certain marginalised segments of the population to the terrible trap that is poverty? What steps must be taken to put a stop to such a process, to prevent poverty by leveraging the insurance systems and compensating for the adverse effects that are provoked? These are controversial questions that go beyond the sole problem of food security and look at the big picture. A lengthy discussion on these matters would stray from the core concerns expressed in this book; suffice to say that such rifts that lead to gross inequality, between the countries of the world as well as within most countries, represent a real threat. And that the consequences go beyond economics and economic development.

Charting a Course of Action

In situations of crisis and localised food availability dilemmas, the answer is clear: humanitarian aid. The World Food Programme (WFP) was created in order to provide appropriate responses in emergency situations. Numerous non-governmental organisations are also extremely active in this domain. As necessary as this solution may be, it does pose a certain number of problems that have now been recognised and documented, and call for a revaluation of the efficiency of emergency aid mechanisms[8]. The problems arise from distortions and adverse effects related to the bulk purchases which are often carried out in order to make transfers, or to the governance of distribution and the risks of funnelling and embezzlement that this causes. These problems must also be taken into account when determining how these mechanisms are implemented. How can we organise supply and distribution to ensure that the benefits trickle down to those most in need, and to this end organ-

thus describe his or her state of being—while the latter represents an individual's different possible combinations of the former. A capability is thus a vector of the modes of functioning expressing the freedom, for an individual, to choose between different living conditions." (E. Monnet. 2007. La théorie des "capabilités" d'Amartya Sen face au problème du relativisme. *Tracés, Revue de Sciences Humaines*, 12: 103–120).

[8] As in other fields, here we must be wary of dogma and ideology. Emergency aid, in particular emergency food aid, is indeed necessary. Its drawbacks, however, should be recognised without rejecting it as a "lethal weapon". In practice, this term is more applicable to long-term aid, which risks making countries overly dependent on donors. But here again, for the sake of pragmatism, we should aim to correct the flaws of the different current aid mechanisms without throwing the baby out with the bath water (as Dambisa Moyo tends to do in her 2009 book *Dead Aid: Why Aid Is Not Working and How There Is a Better Way for Africa*. Farrar, Straus and Giroux, New York).

ise new supply chains to complement existing ones? What are the criteria for generating and allocating aid, products to be distributed and supply chain methods? It is also necessary to monitor aid and make sure it does not occur at the expense of the local reconstruction of institutions and production and sales capacities, which could risk—as we have too often noticed—leading to a long-term cycle of aid and crisis. This is one of the main difficulties facing the implementation of a particular method of territorial development, adapted to unique situations of crisis and reconstruction. With increased prices in foods staples and the intensity and frequency of current crises, amassing global emergency stocks poses a non-negligible financial hurdle.

As we have already seen, hunger is a complex issue that goes beyond emergencies and crises. It is a structural problem with multiple causes, and is constantly evolving over time. It covers a wide variety of geographical situations, from one country or one region to another. These characteristics should prevent us from developing a simplistic, one-size-fits-all solution which would have no real interest or merit, except maybe to serve as a mobilising slogan, but for a limited time only. For example, affirming that hunger can be eradicated through increased production is as simplistic—and ultimately as dishonest—as saying that it is caused by international philanthropy. The same goes for discourses that aim to promote a particular agricultural technology, such as irrigation, GMOs, organic agriculture or a return to traditional knowledge and know-how. Action must be designed and implemented at different levels. The local level, which involves solidarity and income-generating economic planning, is the level at which food access methods must be organised and constructed, but this is not enough. On a national and international level, using an arsenal of adapted policies, it will be necessary to develop strategic orientations for development dynamics, distribution of wealth and land, governance of access to resources, interactions between the environment and the trio of production-processing-distribution, excessive price fluctuations, etc. Whether or not it is agriculture-related, income—from jobs or transfers—is the most important driver of change. In Brazil, the *Fome Zero* project ("zero hunger") is a particularly poignant example of such an arsenal which, during a first phase, focused on the distribution of food baskets to complement retirement benefits and pensions distributed in rural areas. The tools mobilised by these policies—whether they involve only public authorities or integrate the participation of private players—should allow us to manipulate the wheels and drivers of poverty and inequality as well as allow for intervention in times of crisis.

While priority is most often given to urban areas, it is just as necessary—if not more so—to focus on rural areas. The reasons are threefold: first, as we have seen, the majority of the world's poor and malnourished today live in rural areas; second, agriculture and non-agricultural rural activities are likely to have more spillover effects in rural areas; and third, to avoid a mass exodus of people towards cities where they will not find jobs and instead find themselves trapped in a new, urban poverty. This in no way should be taken to mean that solutions must be limited to agricultural policies, nor to increased agricultural production. At this stage, it is interesting to note that urbanisation and the driver of urban demand can, under certain conditions, spur rural, agricultural growth and increased income for rural populations. In this context, regional planning actions will be given priority, especially those that aim to reinforce physical connections (transportation infrastructure) as well as virtual connections (information and communication technologies) between cities and the countryside.

Historically, with the expansion of cash crops (coffee, cacao, tropical fruits, cotton, etc.), access to the far-off markets of industrialised countries played a key role in the growth of the agricultural production of tropical regions, particularly the numerous countries under colonial rule. Over the last few decades, local demand from cities in developing countries experienced a sharp increase and—at least partially—substituted demand from industrialised countries. West Africa is a good example of the spillover effect that urbanisation can have on agricultural growth [14]. One of the major challenges of public policies in developing countries is thus to connect, as much as possible, rural farmers and "regional" urban consumers in order to stimulate food crop production. In order to accomplish this, several steps must be taken: investment in storage and transport infrastructure, regularisation of supply, standardisation of products, implementation of pricing information systems [15], etc. However, the question of jobs remains. On the one hand, connections between urban and rural areas are likely to intensify migratory flows and rural exodus, especially because the competition established between sub-sectors (caused by the application of the dogma of the Washington Consensus during the 1980s and 1990s) has greatly increased the vulnerability of significant populations of rural farmers in tropical countries [16]. On the other hand, and more importantly, the economies of the numerous developing countries must be able to provide jobs, in cities as well as in the countryside, to young people.

In this regard, the lessons of the RuralStruc programme are of particular interest [17]. This programme analyses the consequences of the demographic transformation that is underway or about to occur in the growth of active populations. Job tensions will be the strongest in sub-Saharan Africa, where 65 % of the active population works in agriculture—a percentage that reaches 75–80 % in Sahel countries—and where the demographic transition is still underway (see Fig. 9.1). Some 17 million young people flood the job market here each year (11 million of whom live in rural areas). This number will increase to 25 million in 2025. For an average sub-Saharan country with approximately 15 million inhabitants, this means that 300,000 extra people are arriving on the job market every year and this figure will reach 400,000 per year by 2025. In total, over the next 15 years, sub-Saharan countries must create 330 million jobs, or—to give an idea of the enormity of the task at hand—the equivalent of the current population of the United States. Of this total, 200 million jobs (60 %) must be created in rural areas.

Considering the demographic growth expected in several developing countries, notably in sub-Saharan Africa, we are faced with the very real challenge—and thus the related potential dangers—of rethinking job creation. In particular, we must reevaluate models of agricultural transformation measured in terms of criteria which take job creation into account, and not simply the productivity of the land (the importance of which is not to be underestimated). Given this context, we would do well to highlight, without going into detail, the importance of a framework and

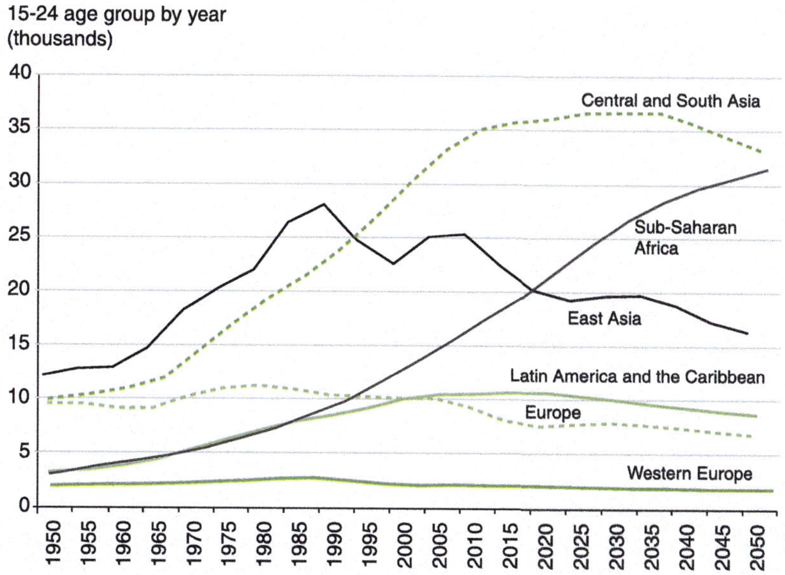

Fig. 9.1 Average annual growth rate of the active population of different world regions through 2050. (Source: United Nations, World Population Prospects: the 2008 Revision, Reproduced with the permission of F. Bourin Publishers)

macroeconomic policies, with one particular watchword in mind: humility[9]. Just as some people see development aid (as it is currently distributed) as useless at best and counterproductive at worst [18, 19], while others see it as more crucial than ever, likewise there is no agreement on which development policies to adopt. Nevertheless, a majority of people tend to challenge the Washington Consensus[10] and point out that there are, and always will be, several paths that lead to growth. Therefore, policies must be varied and adapted to the specific needs of the countries where they apply.

All economies—and those of developing countries are no exception—bow to market prices. This applies to goods, factors of production—most importantly labour and capital—and money. This is not, however, synonymous with unfettered liberalism: while the financial system must be free to take risks, it must also be regulated in order to avoid abuse. The job market must find a balance between the gains in efficiency allowed by a certain flexibility and the gains achieved from a certain level of stability that increases the ability to forecast and ensure job security. The same principle of pragmatism and "custom-made" solutions applies to agricultural and food production markets, and to the regulation thereof. As noted by Dani

[9] This paragraph on development policies is largely based on the dossier dedicated to the issue by the magazine *Sciences humaines* (n° 216, July 2009), which we strongly recommend.

[10] In simplified terms, the Washington Consensus consists in applying the same developmental policy to all countries. The policy is based on a limited number of principles, notably the management of public spending, the non-intervention of the government in the market and thus the privatisation of public enterprises and the financial liberalisation of prices and exchanges.

Rodrik [20], professor in International Political Economics at Harvard University, "rule-of-thumb economics, which has long dominated thinking on growth policies, can be safely discarded". There is just one exception, however, to this principle of model diversification: the key role played by institutions, without which no economy, including the market economy, would be able to function efficiently over the long term.

More specifically, in terms of undernourishment, we must act on different levels and at different scales, while ensuring consistency and favouring comparable data measurement. We would do well to remember that recognising the multifaceted nature of undernourishment is key to charting a course of action that targets all causes of the problem simultaneously, and not just food-related causes. As such, let us highlight three trends in current thinking.

Treating the most serious consequences of undernourishment Medical monitoring of mothers and young children, food education and nutritional supplements in school are all drivers for action with which nutritionists are familiar. The window of 1,000 days, from conception to a child's second birthday, represents the period in which children are in the most need of sufficient balanced food, health monitoring, maternal care, etc. During this period, the impacts of malnutrition, if unchecked or untreated, are often irreversible; they will affect the entire social and productive life of the adult.

Ensuring the availability of sufficient, high-quality food Dynamic and reactive agricultural production is of course a vital condition to ensuring this availability, but this alone is not enough. All the segments of the supply chain which follow initial operations and bring food products to the market also come into play, including harvest, storage, primary and secondary processing, and wholesale and retail distribution. For regions where we know there is and will be a lack of agricultural and agri-food products (namely, Africa and Asia), this will require unrestricted trade, regulation of security stocks and particularly close attention paid to price fluctuations. Agriculture will also be called upon to respond to the issue of nutritional quality of food for undernourished populations and the diversification of this food (fruits and vegetables, dairy products and meat, etc.). For this, local and peri-urban agricultures must also be mobilised.

Improving access to food through household incomes and food prices Fair and sufficiently stable agricultural prices on a local as well as regional and international levels are necessary so that farmers can earn a decent living, boost their investment capacity and generate sales, be it of food or other goods. Costs must also be lowered to ensure access to food by addressing problems of infrastructure and production organisation. Ensuring that farmers receive a decent income in exchange for their work through prices that are sufficiently high without hindering universal access to an adequate, balanced diet is as much the realm of economic development and sharing the fruits of labour as income levels and the distribution of wealth. Beyond these national, macroeconomic issues, which require sound management adapted to the specific needs of each region, excessive fluctuations of agricultural and food prices must also be attenuated. To this end, we will borrow largely from the conclusions of a study carried out by the French Strategic Analysis Centre (*Centre d'analyse stratégique*) [21]. Our recommendations are thus the following: (1) improve the functioning of physical markets of agricultural commodities through a new global governance, the main goals of which should be to prevent crises and provide solutions (i.e. constituting and managing emergency food stocks); (2) forbid policies that restrict exports and are decided in a unilateral, uncoordinated manner; (3) improve the functioning, transparency and regulation of financial markets all while ensuring that they do not contract, because sound management of markets is beneficial; (4) limit the negative impacts of excessive volatility, for farmers (through countercyclical payments, insurance, etc.) as well as for consumers (through public financing of food purchases for the poorest households, social safety nets, etc.).

Food insecurity is first and foremost a question of insufficient development and poverty at national and individual household levels. In order to tackle these problems, solutions must be sought from both within and beyond the agricultural sector. Non agricultural players have an important role to play, be they educational or health institutions, or providers of money or social protection. Agriculture, however, remains paramount, and without it, food security is a losing battle. Firstly because agriculture often represents the only opportunity for the economy of developing countries to take off, especially for the poorest countries, and can provide an excellent return on investment under certain conditions—as the World Bank's World Development Report 2008 showed. Secondly because the poorest—and thus most vulnerable—populations, the "bottom billion", are mostly farmers or rural dwellers. In order to rise to the challenge of global food security, we must thus focus on farmers by giving them the tools to earn sufficient wages from agricultural activities and encourage public policies that foster non-agricultural job creation in rural areas.

References

1. FAO (2010) L'état de l'insécurité alimentaire dans le monde: combattre l'insécurité alimentaire lors des crises prolongées. FAO, Rome, 64 p
2. Delarue J, Mesple-Somps S, Naudet JD, Robilliard AS (2009) Le paradoxe de Sikasso. Coton et pauvreté au Mali. Document de travail DIAL, Paris, Ird, Afd, 29 p
3. Black RE, Allen LH, Bhutta ZA, Caulfield LE, de Onis M, Ezzati M, Mathers C, Rivera J (2008) Maternal and child undernutrition: global and regional exposures and health consequences. Lancet 371:243–360
4. http://ipar.sn/IMG/pdf/RuralStruc_Afrique_diaporama_PTF_2009.pdf
5. http://www.fao.org/docrep/013/i1683f/i1683f.pdf
6. Welt Hunger Hilfe, IFPRI, Concern Worldwide (2010). Global hunger index. The challenge of hunger: focus on the crisis of child undernutrition. Bonn, 54 p
7. Delpeuch F, Maire B, Monnier E (2006) Tous obèses? Dunod, Paris, 188 p, Collection Quai des Sciences
8. http://www.fao.org/fileadmin/user_upload/foodclimate/HLCdocs/declaration-E.pdf
9. Sen AK (1981) Poverty and famines. An essay on entitlement and deprivation. Oxford University Press, USA, 230 p

10. Sen AK (2003) Un nouveau modèle économique: développement, justice, liberté. Odile Jacob, Paris, 368 p (French translation)
11. http://www.fao.org/docrep/005/y4671e/y4671e06.htm#fn21
12. Bricas N, Daviron B (2008) De la hausse des prix au retour du "productionnisme" agricole: les enjeux du Sommet sur la sécurité alimentaire de juin 2008 à Rome. Hérodote 131:31–39
13. Caron P, Schmitt B (2010) Les multiples productions et fonctions de l'agriculture. Dans La question agricole mondiale. Enjeux économiques, sociaux et environnementaux. La Documentation française, Paris (Chap. 8)
14. Bricas N, Thirion MC, Zoungrana B (2009) Bassins de production et de consommation des cultures vivrières en Afrique de l'Ouest et du Centre. Étude Cirad-AFD-Cilss-Ifad, 51 p
15. http://www.oecd.org/dataoecd/25/8/36562896.pdf; et http://www.iied.org/pubs/pdfs/9153IIED.pdf
16. Mazoyer M, Roudart L (1997) Histoire des agricultures du monde, du néolithique à la crise contemporaine. Éditions du Seuil, Paris, 545 p
17. Losch B, Freguin-Gresh S, White E (2010) Structural dimensions of liberalization on agriculture and rural development—a cross-regional analysis on rural change. RuralStruc program, final report, World Bank, Washington D.C.
18. Moyo D (2009) L'aide fatale, les ravages d'une aide inutile et de nouvelles solutions pour l'Afrique. J.C. Lattès, Paris
19. Easterly W (2006) Les pays pauvres sont-ils condamnés à le rester? Editions d'Organisation, Paris (original edition: MIT Press, 2001)
20. Rodrik D (2008) Nations et mondialisation. Les stratégies nationales de développement dans un monde globalisé. La Découverte, Paris (published again in issue 216 of Sciences Humaines). See also one economics, many recipes: globalization, institutions, and economic growth, Princeton Univ. Press
21. CAS (2010) Volatilité des prix des matières premières. Volet 2. Produits agricoles: limiter la volatilité ou en atténuer les effets. Centre d'analyse stratégique, La note d'analyse, janvier 2011, n° 207

Chapter 10
Towards a Global Governance of Food

The importance of food security and nutrition security will continue to grow in the future. These issues, which are a potential source of conflict, will constitute an essential parameter of security itself, whether it is within a country or on a planetary scale. On a more fundamental level, the tragedy of hunger, which affects a large portion of humanity, and the injustice of their inability to access healthy, varied and balanced food, are both tied to fundamental human rights. The 1948 Universal Declaration of Human Rights, and all the international legal instruments that have come after it over the decades, remind us that the rights linked to human dignity are fundamental values, including the right to feed oneself. The planet is a closed system, and any action on one element in a given location affects all the other elements of the system. Because of their dependence on the planet's resources, their links with human activity and human well-being, as well as the diversity of demographic and ecological situations, agriculture and food are a clear example of a system where the actions and strategies of some can greatly affect the lives, health and environment of others. There are similar links, through time, with future generations: the actions and decisions we take today will affect their ability to produce food and feed themselves. This is why reflection and action on food and agricultural systems requires long-term, local and global coordination, if possible cooperation, and preferably solidarity. Finally, it is in the interest of all the players in this system to ensure that it functions as best it can and without any hiccoughs.

Research on agriculture's ability to feed the world has led to extreme, contradictory conclusions. In order to illustrate the diversity—or contradictions—let us take the example of two international conferences, one on organic agriculture which was held in Rome in 2007, the other on climate change held in Copenhagen in 2009. In both cases, it was concluded that agriculture can satisfy global food demand, but the Rome conference concluded that agricultural production must take priority to "target local food needs in local markets; allow import only for items not grown locally; and export high value products", while the Copenhagen conference concluded (based on analyses carried out by the Potsdam Institute in order to propose allocation choices for optimal global land use) that "the 12 billion people with 1995 dietary habits could be nourished on less than one third of the present agricultural area—if the best sites were used for the most appropriate crops and if

M. Guillou, G. Matheron, *The World's Challenge*,

DOI 10.1007/978-94-017-8569-3_10, © Éditions Quæ, 2014

world food trade would operate undistorted by protectionism" and added that "this bold approach would only become feasible, however, if the prime locations would be reclaimed/reserved for agriculture as part of a long-term global deal". Which to choose: localisation of production, or specialisation and intensification of the regions of the planet most favourable to crops? Two contradicting diagnoses from different world views which call for a supported, critical analysis then demand local and global political choices. Because of our conviction and our recognition of the role that research, especially agronomic research, plays and will play in illuminating these questions in an uncertain climate, we have decided to give a critical overview of the situation and express our opinions.

Re-examining the Food Riots

The 2008 food riots generated strong reactions in public opinion. It truly shocked people that this problem could be something other than a simple famine linked to this or that local situation, that it was actually a global scourge. The spectre of food insecurity stayed with people and the entire planet, under the influence of media pressure, quickly—but perhaps only briefly—rallied for the cause. These riots brought the problem of food and agriculture back to the forefront, where it had not been since the 1970s. This reaction was all the more paradoxical because, as we now know, this particular crisis was not calling agricultural production directly into question. We were in fact witnessing the concerted emergence in 37 developing countries of revolts linked to the inability of certain populations to buy basic food products, which were available, but too expensive. These uprisings mostly remained in the urban centres from which they emerged, in close relation to the strong increase in global prices of food commodities over several months. The crisis was linked, first and foremost, to poverty, or the lack of sufficient income for too many households. Even today, crises in the Arab world have found fertile terrain with increased food prices.

Global and national prices for numerous products began to increase in late 2006 and did so even more markedly throughout 2007 [1]. They continued to rise during the first months of 2008 before ebbing dramatically in late spring, only to stabilise at the end of 2008 and the first weeks of 2009. After remaining low throughout 2009 and the first months of 2010, prices once again skyrocketed starting in the summer of 2010 and were still increasing in early 2011. The suddenness and the amount of the hike in agricultural prices in 2007–2008 surprised observers, just as their drop starting in the summer of 2008 did, as well as the new rise starting in the summer of 2010. Studies analysing these episodes of rise and fall have flourished. Unfortunately, there is still no consensus on the respective contributions of the different determining factors causing changes. The increase in global agricultural prices—from the last 6 months of 2006 to the first months of 2008—was the result of a combination of different cyclical factors in the context of a trend that had been going on for at least a decade. This sprang from sluggish supply of agricultural products, dynamic food demand and the consecutive depletion of stocks. Cyclical factors had a negative effect on the quantities available in international markets (e.g. climate incidents and restrictive export policies), a positive effect on global demand (e.g. depreciation of the US dollar, development of first-generation biofuels, need to feed cattle, policies which encourage imports and speculation) and a positive effect on production costs (e.g. increased oil prices and prices of inputs in general).

The sudden drop in international prices of agricultural products, which occurred in the spring of 2008, can be explained by a turnaround, as several factors that had been favouring increased prices were now applying downward pressure: a favourable climate in the various areas of production, the appreciation of the US dollar, the rapid exit of speculators and the very sudden and very significant drop in oil prices. In addition, high agricultural prices in 2006 and 2007 led farmers to increase the surface area of their field crops, to the detriment of fallow land or land previously allocated as grassland, or even forest. They also led farmers to seek the highest yields possible and thus to intensify. And prices picked up again in the summer of 2010 for several reasons: it started with fires in Russia and the country's unilateral decision to limit its exports to favour the domestic market, which was followed by flooding in Australia, the continued rise in the price of a barrel of oil, speculation in agricultural commodities, etc.

What do the recent changes in agricultural prices teach us? There are two main lessons. First, that it is necessary to pursue research that aims to explain the origins of fluctuations, more specifically to quantify the relative weight of the different determining factors of these fluctuations, all the more so because these factors and their relative importance can change from one episode to another. If we do not do this, it will be impossible to develop adapted instruments of price stabilisation and/or correction of the negative effects of their excessive instability. Because—and this is the second lesson—too much fluctuation in agricultural prices is not good for anyone: prices that are too low discourage investment, which is needed more than ever in agriculture, while prices that are too high penalise consumers, especially the poorest among them who spend the largest portion of their meagre revenues on feeding themselves. Concretely, at the beginning of 2010, the number of undernourished people passed the unimaginable threshold of 1 billion people, for the first time in the history of humanity. Although this figure dropped significantly in 2010 (thanks to economic upturn in numerous developing countries), it will most likely increase again (barring any economic miracles) in the near future. It bears repeating that undernourishment—of which food riots are only the most visible facet by being the most mediatised—is first and foremost a question of access to food, more precisely a question of difficult, sometimes impossible access because income is too low or because food prices are too high (see Chap. 9). By 2050, global food demand should increase strongly because of demographic growth and evolving diets. While the World Bank's World Development Report 2008, which was dedicated to agriculture for the first time in 25 years [2], and the IAASTD Global Report [3] from the 2008 Session were only just published (without having anticipated the 2008 crisis), numerous initiatives have materialised since that year. Whether focusing on food distribution, support for household income and access to food, policies that limit fluctuating food and agricultural prices or reinforced production capacities in the agricultural sector, they anticipated the political mobilisation that spread across the planet starting in 2008, even if the concrete effects remain unclear at the time of writing this book.

We Must Step Back From Agriculture in Order to Understand It

"The world is hungry." This surprising catchphrase, conveyed and amplified by the media, was used to interpret the 2008 food riots, which occurred in the capital cities of 37 countries. We were assured that the answer to these riots was to produce more. Their impact was thus to create a new vision where development in the agricultural sector was key. However, more in-depth analyses—particularly those carried out in the research community, in international bodies like the World Bank or in non-governmental organisations such as Oxfam—quickly showed that this was overly simplistic, and that what was in fact at play were the means of regulating consumption, production and trade of agricultural and agri-food products.

We must distinguish the specific problem of the 2008 riots from the long-term challenges of being able to feed the planet in 2050, within a sustainable agricultural and food system, and bringing the "bottom billion" above the poverty line. These two challenges are real. The second, the solution to which can only be imagined in the long term, covers the following important questions: How can we ensure the efficient and transparent functioning of the markets and secure trade? What models of development will be needed for this? And how will they impact the process of impoverishment? We cannot be satisfied with a world where one part of the planet feeds all the inhabitants to the detriment of environmental and social functions played by agriculture in the other regions. Agriculture plays a key role in a large number of environmental issues, whether it is the conservation of natural resources or their contribution to the management of local or global environmental goods, including landscapes. It is at the heart of the development process and thus of the economic, social and political security of the so-called developing world. And what is true for these countries is true for many industrialised countries. In many regions of the world, the future is being determined through the creation and distribution of wealth, and through the capacity of agricultural activity to generate jobs—whether directly or indirectly—which is a very important agricultural output. Agriculture is also at the heart of poverty issues, not only in rural areas—home to a large majority of people suffering from hunger—but also in urban areas, through the connections and migratory flows that exist between the two spaces.

If we wish to understand food, agriculture and their interactions, we must extract them from their narrow, traditional fields. Climate, economic growth, currency parities, energy prices, diets, sales strategies for distribution and even the intervention of operators on the commodities markets (including speculation) are all factors to be taken into account in order to better understand the evolution of supply, demand, trade and, finally, necessary foodstuffs. The studies carried out by our two institutions, INRA and CIRAD, provide potential directions. For example, the rarity of fossil fuels could become the main constraint for the evolution of agriculture in the world, which must use these inputs more efficiently, both when used directly or indirectly in the form of an input for fertilisers or concentrated feed. At the other end of the spectrum, the demand for plant-based fuels, chemicals and materials will be influenced by reserves of oil and gas, their consumption and their prices. In the future, prices of agricultural resources will be even more tied to those of fossil fuels: fluctuations in oil prices will increase the instability of the prices of agricultural products.

Agriculture and food depend on numerous external factors, such as the rarity of water resources, climate and demographic change, economic growth, urbanisation and changing lifestyles. The evolutions at hand in the agricultural and food sectors will, in turn, impact these external factors—agriculture and food can no longer be understood and analysed without keeping this circle in mind. In this perspective, one of the goals of the Agrimonde platform—which we wish to establish as a permanent feature—is to specify the potentially limiting role of these external factors, which will be necessary to the sustainability of agriculture and food, and the effects these two activities will have on these limiting factors.

Diets: A Factor of Future Need

In this book, we have described the different transformations that agricultural systems will have to undertake in order to produce more in a more ecological manner, and also in order to produce other items, be they bioenergies, biomaterials or environmental and territorial services. These transformations will require new knowledge, as well as an increased capacity to integrate, invent and innovate. In any case, the viability of these new agricultural systems must be evaluated, in developed as well as developing countries, not only in terms of economic criteria, but also on the basis of environmental and social criteria. By making a conscious choice to work towards sustainability, we must optimise all the processes, safeguard the potential of ecosystems over the long term and use inputs as efficiently as possible, given their finiteness and the productivity we wish to achieve. These same principles must be applied to food systems, to consumer behaviour and to the strategies of companies that transform and distribute agri-food products.

While biofuels are often blamed for shortages in analyses of the current and future food situation, abuses and waste in certain systems of food consumption are often forgotten. People's diets are the most important factor of demand, both in terms of quantity and quality. Losses and waste in distribution and consumption circuits also play a large role in industrialised countries. Public and private players must be informed of this situation and identify the drivers that each one can employ, independently and conjointly. The collective scientific expert report recently compiled by INRA at the request of the French Ministry of Food and Agriculture [4] showed that questions of lifestyle, urbanisation, labelling and portion sizes play an important role in consumer behaviour, in addition to income, prices, food composition and even marketing. It is now not only possible, but also desirable, for public policies to play a role in this field essential to health and the sustainability of planetary food systems. Overweight and obesity are no longer only a problem in developed countries; they also affect emerging and developing countries alongside nutritional deficits and undernourishment: this is the "twofold challenge" with which more and more countries are now faced.

Today, according to the WHO, 65 % of the world's population live in a country where overweight and obesity are a bigger (no pun intended) problem than undernourishment. On a global scale, nutritional issues are insufficiently covered because the organisations in charge of food (FAO) and health (WHO) are distinct and disconnected. It is of the utmost importance that we work toward a global agenda of nutrition security. The innovative Agrimonde 1 scenario from the Agrimonde foresight study took this into account, by focusing on a reasonable diet in all regions of the world of an average 3,000 kcal per day per person, with an average 500 from animal products. There are top-priority actions to be carried out on three levels. First, there needs to be more investment in research in order to both better understand food consumption patterns in all their dimensions—especially behavioural, social and economic—and the drivers for action in order to modify them and anticipate global food issues within a balanced system of interactions between supply and demand. Second, it will be necessary to develop and implement public policies which favour healthier, more sustainable eating habits, both in the long term (through education and training) and the short term (through positive and negative tax incentives). Third, we must tackle this problem of losses and waste head on, with the double perspective of reducing and recycling them.

Reinvesting in Agriculture: An Urgent Necessity

Many of agriculture's functions are currently being mobilised, both in developed countries and in emerging and less developed countries. Agriculture goes beyond producing food, fibres and energy; it is also a powerful driver for rural development. A source of wealth and job creation, it affects resources and people's health, as well as the environment, territories and landscapes. Agriculture is so important that, after a long period of ignoring it, we now risk overdoing it and asking too much of agriculture and farmers, without really knowing how to "compensate" the latter for the services they provide to society. When we see the portion of state budgets dedicated to agriculture, the paradox is obvious: countries that depend most on agriculture for their growth spend very little—less and less in fact—on this sector. Only 4 % of GDP is dedicated to agriculture in sub-Saharan Africa, while 10 % would be needed for sustainable growth [2]. What makes matters worse is that, from this meagre total, most of the investment goes towards urban rather than rural areas. The urban bias can be partly explained by less media attention and the weaker political force of rural populations. Hunger in cities motivates public opinion and politicians more than hunger in the countryside, even though worldwide hunger is a mostly rural issue. As for international development donors, the proportion of public aid going to rural areas has already greatly decreased, from 20 % to less than 5 % in just 20 years. The amount and distribution circuits of this aid are particularly opaque, to the point where certain players wonder if it is being done on purpose and if there may not be a "culture of not fulfilling commitments" on the part of donating countries, who, summit after summit, accumulate promises they never keep, despite their charming displays of compassion.

The impact this aid really has is a point of controversy, because of the modalities of its in-field mobilisation, as well as the dependency that it can sometimes provoke. Oft-changing leadership reduces our ability to evaluate these situations to relying on indicators which are sometimes divorced from reality. As for the specific case of French aid, the strategic choice to favour concessional loans, by strongly reducing donations, excludes both the least developed countries and the agricultural sector. By doing this, support for agricultural policies is not available to the countries which need it the most. While investments by OECD countries in Africa (all sectors combined) have decreased, those by developing countries tripled between 2004 and 2008 and now represent more than a third of total investment in Africa. UNCTAD's recent Economic Development in Africa Report [5], dedicated to South-South cooperation, highlighted issues such as the explosion of Chinese financing of African infrastructure (from 470 million US$ in 2001 to 4.5 billion US$ in 2007) and points out that the agricultural sector should be able to benefit more from this situation.

Taken together, these evolutions urge us to profoundly re-examine not only the amount of aid and credit allocated to the agricultural sector, but also top-priority targets and modalities for aid distribution. But in the same way that food security cannot be simplified as a reflection of production, reinvesting in agriculture cannot be limited to a simple quantitative equation of agricultural production. It requires the support of a certain vision of development. The demographic evolution of Africa means that today around 17 million young people arrive on the job market each year; ensuring work for these young people is an absolute priority for the continent. The non-agricultural and non-rural economic sectors do not seem to be capable of supporting these jobs. Agriculture and rural jobs will continue to play a pivotal role in African economies for the next several decades. On the other continents, the terms of the equation may not be quite so strict, but the family farming sector still represents a significant source of jobs and of considerably increased production, if intensive agriculture is used intelligently. A large number of the planet's inhabitants are peasants and many would like to remain so, but we must make sure that this is not their only option.

Investment strategies can no longer be based on the dual, exclusive vision that was dominant for so long (that of a capitalistic, "modern" agriculture versus a family-run, "archaic" agriculture), nor be limited to advocating only the development of heavy investment in "modern" agriculture. The example of Brazil—poised to become the world's leading agricultural producer, despite a population which is 85 % urban, with 600,000 company operators and more than 6 million family operators—is striking: in 2005, 32% of Brazil's agricultural GDP was generated by family farms that accounted for 80% of the domestic production of tubers, more than 60% for beans, 40% for rice and maize, etc. This family-based agriculture made significant contributions to Brazil's agricultural trade surplus, while only receiving 37% of funding from public financing [6, 7]. Outside this large agricultural country, the organisation of farmers (including livestock farmers) and their capacity to innovate and to access financing—as modest and parsimonious as many microcredits analysed by Esther Duflo may be [8]—will all contribute to the drivers for agricultural development, and thus rural and social development. The fate of more than 3 billion people hangs in the balance, whether they depend directly or indirectly on farming for their living. Nevertheless, they must be part of innovation, markets and economic and human development. These are major issues and carry a certain risk. All loans and investments, particularly those coming from public aid, must take into account these top priorities which require the triple foresight of a productive, ecological agriculture which responds to the social need for job creation.

Fighting Poverty

The food emergency is now essentially a question of consumers' ability to pay, and thus a question of poverty. In most predominantly agricultural developing countries, there is a strong link between increased agricultural production and poverty reduction. Agricultural development depends greatly on agricultural and rural investments. Irrigation, land improvement, storage facilities for factors of production and products, energy, roads and communications all contribute—whether directly or indirectly—to the development of agriculture. Global diagnoses alone cannot help develop the solutions to be implemented. The large diversity of situations requires differentiated responses. Known as "donor darlings", certain countries have had real success stories in agricultural and rural development, including Ghana, Malawi and Mali. In Mali, investments in agricultural productivity generated higher incomes in both rural and urban populations. By lowering food prices and increasing purchasing power, these investments helped improve diets in the different types of households studied. The dilemma of choosing between helping urban or rural populations is thus nothing more than a myth. We must remain wary of sweeping generalisations and take into account local specificities which shape the efficiency of investment and the reach of its positive impact.

Rural development depends first and foremost on basic investments in areas such as education and training, health and water sanitation, which all have a direct impact on quality of life and without which additional investments in other areas would not be possible. At first, these should be financed through public spending, without excluding private financing that stable countries could attract. Prioritising these types of investments between rural areas and urban demand is often difficult. Once again, the lack of political and media attention often penalises rural populations. There are also non-material investments to be made, in terms of public goods which benefit everyone and are indispensable in order to pave the way for other, more risky investments. These are the responsibility of individual countries. Besides civil peace, which unfortunately does not exist everywhere, there is what we call "good governance", i.e. the functioning of institutions, loan distribution circuits, regulatory transparency, etc. These are all important factors for attracting other resources, mainly because they represent security and transparency. The existence of these contextual elements is not a given, as we have seen by examining the sometimes precarious status of land ownership in many countries.

Even though the Millennium Development Goals—solemnly adopted by the United Nations in order to reduce poverty by 2015—are monitored from year to year, the results obtained are not very encouraging. They even declined because of the 2008 food crisis, and then the 2009 economic crisis. They did, however, improve in 2010, but are likely to regress once again in 2011. Experts agree that increasing investments alone will not help achieve these goals. We must also radically redefine the way we approach the issue and the sectors on which we must focus. According to more than 700 case studies, investment in agriculture, or more broadly the rural sector, is the best driver in the fight against poverty; it is clearly the sector that currently requires the most long-term, intelligent investment in order to reduce poverty and increase food security and peace throughout the world. Because the financial and economic crisis has limited the ability of states to intervene, at least for developed nations, we must encourage private players to put their resources to use. These material and non-material investments, from public and private sources, will only be efficient if they are carried out in a stable macroeconomic environment. To this end, all institutions (education, justice, health, etc.) must operate transparently and efficiently, with no exception.

Regulation: Rules as a Safety Net

The 2008 crisis unfortunately showed how lacking the regulatory systems of agricultural and food products are, whether in terms of price instability, storage capacity, collaboration between the different players, market rules or risk. An African household that spends 60 % of its meagre income on food cannot withstand an increase in food prices of 30, 40 or 60 % without being confronted by hunger and then revolt. Relative price stability is widely recognised as being crucial, for both consumers and producers. However, there is no consensus on the role to be played by public regulation and the tools needed. Volatile agricultural prices have become the norm. They will most likely become even more volatile in the future, under the twofold weight of climate change—which will multiply extreme weather events—and the financialisation of agricultural markets, which could amplify variations (both increases and decreases) of the prices of agricultural commodities, as well as their rate of variation. It is in everyone's best interest to limit the excessive price fluctuations of agricultural production, which penalise both extremes of the supply chain: too high, they hurt poor populations; too low, they hurt farmers who can no longer invest.

Besides stabilising these aspects, it will be necessary at the same time to secure the trade of agricultural and agri-food products. The prospective medium- and long-term scenarios, which take into account the conditions of climate change and demographic trends, all highlight the growing need for these exchanges. Even in very optimistic scenarios that foresee increased food crop production, such as the Agrimonde 1 scenario from the study of the same name, many important regions of the world will still have a food deficit. Africa, Asia and the Near and Middle East will need to import agricultural and food staples from areas with surpluses, such as Latin America (particularly Argentina and Brazil), North America, Western Europe and the countries of the former Soviet Union. In order to ensure that the countries dependent on international and regional markets for their food do not suffer from shortages, trade must be secured. This is especially important because the first victims of supply restrictions will be the most vulnerable populations.

As we have mentioned, rich countries do not agree on the regulation of prices and agricultural exchanges. Research on the subject has not provided solutions that everyone accepts. This is because former experiments, for example in terms of food storage or price regulation, have often failed. However, this negative assessment should not discourage us from taking action because there is a real need to regulate the trade of agricultural and agri-food products. There are four main priorities. First, we must forbid unilateral policies which ban or even restrict trade. Second, we must recognise that the current framework of the WTO is not sufficient. This does not mean returning to protectionism, but total market freedom of agricultural exchanges is not the answer. Giving the poorest countries special treatment (applying common rules, with a longer period of implementation) will not be enough. Third, economic, environmental and social aspects must be better linked (in terms of multilateral trade discussions). And fourth, we must provide the poorest countries—who in any case carry little weight in agricultural trade—the proper conditions to jump-start their economies, using agriculture, which is very often their only source of wealth.

In order to prevent excessive fluctuations of agricultural prices and the negative effects these have, investments must be made in developing countries' agricultural systems so that these countries are less dependent upon imports and thus less susceptible to sudden price variations. Investments must also be made in agricultural production throughout the world in order to increase global supply of agricultural products and sustain the increase in demand which is very likely to take place. Climate events in important production areas will doubtlessly become more frequent because of global environmental changes, while food consumption will increase. It is therefore our opinion that there must be global transparency for the information about availability and an implementation of global stocks, despite the apprehensions of the many players who put their faith in markets and the stabilising effects of these markets. These stocks would be both virtual (via an information system about public and private reserves and the ability to monitor them) and real to a moderate extent. By making it possible to respond in an emergency situation, these stocks would be used to intervene when global markets needed moderation, or as a tool to fight famine. The G20 meetings are an opportunity for concerted action and the implementation of these preventative and emergency tools on a worldwide basis. We can only hope that the members of this group are conscious of the importance of global solidarity in terms of food and that they will take an ideological stance and find the means to act.

The Need for Global Governance of Food Security as a "Public Good"

Because food and agriculture are not run-of-the-mill sectors, and because they affect our security, our well-being, our environment and the life of our territories and our crops, they require proactive public policies. They require policies that protect the poorest consumers and cushion risk taken by farmers. Governments must arbitrate between the interests of consumers and those of farmers. Moreover, they must reconcile short-term needs and long-term challenges, while ensuring that food and agriculture are positioned at the crossroads of demographic, societal and environmental issues. This cannot occur without new commitment from public authorities, in a way that ensures conditions for greater, better-quality production and food and ensures that trade occurs more easily and under more secure conditions. This will require collaboration and multilateral agreements which make it possible to reimagine and support trade while recognising food security as a major global political goal, and supplying countries in extreme crisis as a shared necessity.

The 2008 food crisis did not only point out the shortcomings of our systems of price regulation for food products; it also showed that there is a profound lack of governance and coordination between the different international players, be they public or private. Because it calls into question the planet's social and environmental equilibriums and because it can lead to crises and tragedies—both locally and globally—food security is much too important an issue to not take centre stage in terms of policy planning. Given the characteristics and the changes in eating habits, production and trade, it is clear that food independence everywhere all the time is a fallacy. Food production is now a part of global trade. Markets are not only interconnected by climate or sanitary constraints which seasonally affect harvests, but also by the decisions, standards and policies taken at a national and international level. This reinforces all the more the all-important necessity of ensuring food security through appropriate, coordinated methods on a worldwide scale.

We must therefore first point out that the right to food security is like a global public good, even though the term may be cause for debate. Situations of extreme poverty and food insecurity are major sources of political instability, problematic migrations and conflict. The race for resources needed for agricultural production, as shown dramatically by the race for arable land, will increase as awareness of their finiteness grows. Unless we are willing to believe that we should all live behind fortress walls, we must admit that it is in our general interest that everyone has a decent diet and the resources necessary in order to produce food. It is thus important to define the terms of a concerted, articulated, global governance of agriculture and food which would allow us to not only face the food challenge in all its different forms, but also to anticipate instability and crises in order to act on them and compensate for them. Besides acting on the volatility of prices and storage, this governance would be needed in order to remunerate environmental services, implement security and quality standards and develop a framework for the financialisation of agricultural products or even for competition between food and energy.

Following the example of the scientific community's mobilisation for the climate and the true originality—despite what critics may say—of the global collective expertise of the IPCC, a similar organisation for biodiversity, the IPBES, has just been officially launched by the United Nations General Assembly. The idea is that to each global public good corresponds a worldwide mobilisation of experts who can provide information to take concerted, consistent public decisions on different levels. Policymakers have begun to discuss certain issues in order to promote the idea of global governance. During the Rome summit on global food governance in June 2008, French President Nicolas Sarkozy launched the idea of a global partnership for agriculture and food which was taken up again during the G8 summit, and continued in other multilateral discussions: in Madrid in early 2009, in Rome in late 2009, then in Aquila during the G20 and G8 meetings, where figures were published. If this partnership is still in the discussion stage, it is doubtless because of the enormity of the task at hand and the difficulty of electing a leadership for such a multilateral subject, but these initiatives nonetheless put forward food security as a global public good. One of the three essential elements recognised by this partnership is that of a network of high-level experts, implemented at the end of 2010, which makes it possible to organise available knowledge and to identify the new priorities and controversies to explore. Global agronomic research should focus on characterising the complexity of food security and in establishing an agenda to meet this challenge.

Dedicated Agronomic Research

In order for this transformation to succeed, and because agriculture has become a "knowledge-intensive" activity, we will need to invest heavily in innovation, particularly through scientific methods and the transmission of knowledge. When there are budgetary constraints, investment in research is often questioned or a high level of return or impact demanded. It is difficult to establish direct causalities between the production of knowledge and the evolution of social and economic realities. However, numerous meta-analyses show a return on investment of nearly ten to one for money invested in agronomic and agricultural research, which clearly shows the efficiency of this type of investment. Why then is this research often under-funded, while only public investment can be used in the fields relating to public goods?

The financial resources of numerous developing countries are lacking and, while the need for international financial solidarity is apparent, in practice the promises made by the governments of developed countries are not monitored closely enough. International solidarity becomes limited when nations are confronted with internal budget problems. It will be important to prioritise the issues to be addressed by research in international institutions, mechanisms and networks, and to ensure that multilateral, financial or large private foundations take an interest in the problems of developing nations. The research community is currently undergoing a profound transformation. New players are emerging: the BRICS (Brazil, Russia, India, China and South Africa) and emerging countries now have powerful institutions for agronomic research, as if in response to the major issue of food and the importance that they attribute to agriculture in their own development processes. The Consultative Group on International Agricultural Research (CGIAR) is undergoing a dramatic transformation under donor pressure and is developing a new agenda. The institutions of OECD countries are beginning a new form of competition, where ownership of living matter, like the race for scientific publications, now plays a major role. In addition, key players from the private sector, sometimes with a global reach, have allocated considerable research budgets.

The analysis of investment in global agronomic research [9] shows that, while the vast majority of the total investment of 23 billion US$ (in 2001) still comes from OCED countries, the situation is changing rapidly, with more and more money coming from emerging countries. Countries with intermediate revenues and poor countries represent 32 and 11 % of these research budgets, respectively. These are merely averages which hide the diversity that exists: certain poor countries have seen their research capacities nearly disappear over the last 30 years, while others have maintained viable research systems, by making it a priority, despite whatever else may have happened. New alliances and forms of collaboration have therefore been created, as well as new forms of competition. Three questions thus spring to mind. First, what is the status of the knowledge produced, particularly in terms of how it is made accessible and disseminated? Second, what forms of international governance will allow for a fluid system of apprenticeship, scientific production and cooperation, and conflict resolution? And third, what will be the link between development and scientific culture? In addition to stimulation from competition, the local capacity for knowledge production has now been shown to be a key factor in the development of economies and societies. The goal of organisations such as ours is to help reinforce these shared capacities of knowledge production.

To conclude, it is the intertwined destinies of agriculture and food and the complexity of their interactions with the broader dimensions of economic, social and environmental development that make it necessary for there to be new regulations and governance, both on a global scale and on smaller, local and regional scales. Researchers are needed to provide their expertise. They must participate by formulating and analysing the new issues that the implementation of this governance will raise, and by contributing to the exercises of forecasting, expertise and studies that are shared on a global level. In agriculture, as in many other sectors, the linear system of innovation is no longer valid. Today, each farmer—be it a cereal farmer in France's Beauce region, a livestock farmer on the Amazonian border or a small cacao farmer in Ghana—is the most important player involved in applying innovation to each context, operation and environment. Agriculture is not—and doubtlessly never will be—a succession of standardised acts and procedures destined for production. The days of "technological packets"—a kit of "ready-made" techniques—are over because the diversity of situations calls for "custom-made" solutions. Farmers will also have to integrate the range of positive and negative interactions between agriculture and the environment into their skills. They will be surrounded by professional consultants specialised in profound transformations who must be capable, at the same time, of preserving and transmitting the enormous capital of knowledge and know-how, and to anticipate the numerous transformations to come. Engineering and its ability to combine knowledge and technology will most likely be a determining factor in the training programmes of tomorrow. We firmly believe that it will be possible to feed the 9 billion people in 2050 and beyond—if, and only if, each of us takes action today. From governments to consumers and citizens, from research organisations to professionals, we all have a part to play.

References

1. Guyomard H (2008) Expliquer les évolutions des cours des matières premières agricoles: à l'impossible, nul n'est tenu! OCL 15(6):364–377
2. Banque mondiale (2007) Rapport 2008 sur le développement dans le monde: l'agriculture au service du développement. Banque mondiale, Washington D.C., 386 p (published on 19 Oct 2007)
3. IAASTD (International Assessment of Agricultural Knowledge, Science and Technology for Development) (2005) Agriculture at a crossroads. Rapport global (a summary exists in several languages for decision-makers)
4. Etiévant P, Bellisle F, Dallongeville J, Donnars C, Etilé F, Guichard E, Padilla M, Romon-Rousseaux M, Sabbagh C, Tibi A (eds) (2010) Les comportements alimentaires. Quels en sont les déterminants? Quelles actions, pour quels effets? Expertise scientifique collective, synthèse du rapport, Inra (France), 64 p
5. CNUCED (2010) Economic development in Africa. Report 2010. South-South cooperation: Africa and the new forms of development partnerships. United Nations, New York, 116 p
6. Estatisticas do meio rural (2008) Departamento Intersindical de Estatistica e Estudos Socio-econômicos, Núcleo de Estudos Agrários e Desenvolvimento Rural. MDA, DIEESE, Brasilia, 280 p
7. Sabourin E (2008) Paysans du Brésil: entre échange marchand et réciprocité. Éditions Quae, Paris, 240 p
8. Duflo E (2010) Lutter contre la pauvreté; Tome 1: Le développement humain; Tome 2: La politique de l'autonomie. Seuil, La République des idées, Paris
9. Beintema NM, Stads GJ (2008) Measuring agricultural research investments: a revised global picture. ASTI Background Note

Chapter 11
Conclusion Preventing Catastrophe

Every catastrophe triggers an onslaught of destruction. And each time, the world wonders what could have been done in order to avoid or limit the damage. Indeed, famines and food shortages have marked the course of human history, but things took a turn in the twentieth century, when the planet was finally producing enough food to feed humanity, despite it being more populated than ever. A stream of alarmist analyses (inspired by Malthus) began to emerge, citing our inability to feed the world, but they have thus far proved false. Hunger is now mainly tied to problems of access to and distribution of food. The current causes of hunger are crisis, war, inequality and poverty. Throughout this book, we have explained the context of the issue and found that if care is not taken, the world's nutritional situation could once again be catastrophic in the twenty-first century. Ensuring food security remains a major challenge. Three foreseeable processes may well converge to cause a catastrophe. First, the population of the planet is at an all-time high, which in turn will lead to rapidly growing agricultural demand. Second is the environment. "Nature", which is now very dependent upon and shaped by human practices, must be managed as a system and collective good with limited resources and as a precious inheritance to be conserved for future generations, even though fossil energy sources are becoming increasingly rare. Third, climate change will heighten uncertainty and lead to critical situations because of the unfortunate convergence of extreme events, such as droughts, flooding or fires.

These global challenges will be easier to face if inequalities between the world's different regions are reduced. Agriculture is an essential driver of development and should allow many world regions to emerge from extreme poverty. People are progressively becoming aware of agriculture as not only a sector that provides food commodities, but also as part of a complex system of inter-related fields (e.g. health, energy, jobs, wealth, political stability, local and global environments, etc.) which fluctuate between crises, degradations and contributions to sustainable development. Agriculture is thus at the heart of major global balances, and is subject to all the tensions and vicious circles which bring about instability, while playing a role in preventing the worst-case scenario. So how is a global catastrophe to be prevented? Certainly not by staying the course. There must be a radical change in

M. Guillou, G. Matheron, *The World's Challenge*,
DOI 10.1007/978-94-017-8569-3_11,

habits, both individual and collective. In order to achieve this, tomorrow's solutions must be different from today's practices.

- Climate and natural resources are now greatly impacted by human activity. The sustainable management of these "public" goods is of the utmost importance. It can only be carried out through a better understanding of the functions and interactions on both local and global scales, through shared knowledge, diagnosis, the adoption of common goals and the implementation of the means to achieve them. Agriculture, forests and rural spaces appear to be more efficient and cost-effective than other means of reducing greenhouse gas emissions. Their future must be conceived with this objective in mind, while integrating innovations, regulatory mechanisms and economic incentives.
- Food systems are key to the basic human right to food. As the Agrimonde foresight study and its two contrasting scenarios show, the demand made on agriculture—and thus global pressures—will differ dramatically depending on the average food calorie intake of the world's inhabitants. Sustainability therefore requires a change in current food and nutritional trends. It is imperative that we develop more balanced food systems adapted to local situations, the implications of which will preserve the stability of global balances. We must "think globally, act locally", to quote the expression used by agronomist René Dubos during the first Earth Summit in Stockholm in 1972. We must also "act globally" because, as explained throughout this book, action must be taken at all levels, not least of all at the world level, which is also the most complex.
- Agriculture is unlike any other activity. Beyond its nourishing, economic and environmental functions, it is also recognised as the main driver for development and the fight against poverty in the all-too-numerous regions of the world which are currently suffering from hunger and whose only wealth is their land and people. These rural populations, who represent nearly one third of humanity, will not all be able to find decent jobs in urban centres—far from it. Their future depends on the economic development of rural areas where the priority must be to create jobs firmly based in territories, particularly jobs in agriculture. It is with these rural populations, who are the custodians of farming knowledge, that we must invent new, global practices and develop innovative agricultural systems adapted to both local conditions and planetary issues. Once again, we must "think globally and locally, act globally and locally".

As with the first green revolution, science will play a key role in providing understanding and innovation, particularly in terms of biotechnology and agroecology, but in an entirely different way. Innovation will come about through collaboration with farmers, their organisations, industry and other partners, fuelled by local practices and situations. Public research organisations will also have an important role to play in order to pool experience—as the Intergovernmental Panel on Climate Change (IPCC) has done for several years in the field of climate change and as the recently created Intergovernmental Platform on Biodiversity & Ecosystem Services (IPBES) should do in the field of biodiversity—but also to openly share genetic resources and knowledge.

Throughout the different chapters of this book, the excessive food consumption of rich countries and households and our responsibility to reduce losses and waste have been investigated; the developing need for biomass in order to partially replace oil and coal has been explored; and the importance of ecosystem services and our connection to the planet has been reaffirmed. In this overcrowded world where cities will be more populated than the countryside, we have endeavoured to explore the solutions brought by reinforcing agriculture in developing countries, but without counting on expanding the world's cultivated surfaces. The focus has been on the mobilisation of our research institutions to help develop diversified agricultural systems, which will be intensive but still sustainable, technologically advanced and adapted to their context. The need to reinvest in agriculture, fight poverty and construct a system of global governance for food and nutrition security has also been reiterated. Now that the urgency of action has been made known, an ethics-driven point of view must be adopted. For Louis Schweitzer, ethics requires us to reflect on what we believe we know. What recommendations, then, should be put up for discussion at the end of this global overview? We propose six. As heads of public research organisations, the first recommendations concern us directly; others concern our leaders, but all six call upon the individual citizen. In addition to being aware of the issue of food security (which is becoming a shared value), each individual should feel that he or she has a role to play.

Our **first recommendation** is to step up biotechnological, agroecological, economic, social and human research to contribute to a dramatic boost in knowledge and allow for useful innovations to be developed and distributed to rural populations throughout the world. The goals we must pursue and attain include: more robust varieties of crop species such as wheat, maize or rice—the staples of diets throughout the world—selected in order to be more productive and require less water, fertilisers and pesticides; livestock farming that is low in greenhouse gas emissions; systems that are resistant to drought or flooding, as well as extreme temperatures; the stimulation of soil microbial life and the prevention of soil degradation. For this, we must remain open to the diversity of paths of investigation, from the stimulation of natural defences in organic agriculture to the use of genetic engineering.

Changing research practices through broadened partnerships is our **second recommendation.** We must dialogue with all players before selecting priorities for action, examine the impacts of innovation on lifestyles or on the distribution of benefits, and monitor training and the sharing of innovations. Work must be done in all the fields related to global public goods such as climate, biodiversity, water or soil, within international networks in order to pool data and contribute regularly to the sharing of achievements and uncertainties to inform decision-makers. A global data collection, integration and modelling platform (such as exists in the energy sector) must be developed for food and diet.

Our **third recommendation** concerns the fight against poverty and hunger. Here, the main obstacles are economic. To overcome them, material and human investments must be made in agricultural supply chains throughout the world in order to provide an economic driver and act as an essential source of wealth in rural regions which are home to three-quarters of the world's undernourished people. Fighting losses in the agricultural systems of poor countries, in the field due to disease and plant and animal parasites, and in storage and transport will require investment, for rural populations and for their organisations as well as the processing sector.

Our **fourth recommendation** is to minimise excessive price fluctuations of basic food commodities. Such volatility aggravates poverty at both extremes, bankrupting vulnerable farmers when prices are too low and starving poor populations in cities and the countryside—as well as "landless" rural populations—when they are too high. The implementation of different mechanisms on the part of global policy-makers—emergency stocks to be used in case of famine or excessively high prices, rules of transparency in order to better inform operators in all markets, rules limiting speculation on agricultural commodities—will be necessary if we wish to limit excessive volatility of the prices.

Our **fifth recommendation** involves food systems considered in light of the inextricable challenges of changing diets that cause overnourishment and overweight, reducing food losses and waste and implementing incentives for eating habits that respect the environment and natural resources. In order to have a significant, lasting effect on eating habits, we will need—in addition to education, training and public nutritional recommendation campaigns—regulation of labelling, advertising, catering and urban planning.

Our **sixth recommendation** is to develop a system of global governance for food security. Currently divided between several organisations, this governance must be rethought, reconstructed and reinforced. It must allow us to develop legitimate mechanisms through which actions will be decided on, financed and implemented.

To sum up, the global agricultural system cannot go on with no one at the helm. We must have solidarity on a global scale: environmental solidarity because of the proven influence of man's actions on climate change, the importance of preserving biodiversity and the depletion of natural resources, which generally cannot be compensated for with man-made substitutes; food solidarity in times of emergency, but also in order to secure global trade and avoid excessive price fluctuations of basic food commodities; intellectual solidarity in order to better understand, diagnose and innovate through the mobilisation of national and international agronomic research and development organisations throughout the world. Action must be taken now to shape a better tomorrow. As André Gide once said, "It is not only the world that must change, but man [1]".

References

1. Gide A (1935) Les Nouvelles nourritures. Editions Gallimard, Paris

Glossary

Agroecology Agroecology is a science, an agricultural practice and a movement. As a science, agroecology aims to combine ecology and agronomy (which includes animal production science) with economic and social sciences and apply those principles to cultivated land. With regards to agricultural practices, agroecology focuses on the preservation of the environment and natural resources to take full advantage of the services provided by nature. As a movement, agroecology is opposed to industrial agriculture, which requires production factors from outside the farm. Supporters take issue with a purely technical approach to agriculture and advocate a global approach based on the value of traditional practices and know-how. No standard definition has yet been established for any of the three aspects.

Allele Alleles are different possible versions of a single gene. Each allele is one of many differences in the DNA sequence which holds genetic code (chain of nucleotides in four different forms—A, C, G and T—equivalent to a four-letter alphabet).

Bale wrapping A conservation method that consists in wrapping a plastic sheet around individual hay bales during baling to keep out air and promote lactic fermentation. It is only useful when the hay moisture content has been highly reduced by partial curing. Curing consists in cutting the grass and letting it dry.

Biodegradability The ability of a substance to be broken down by micro-organisms such as bacteria, fungi or microalgae.

Biodiversity The natural diversity of living organisms. Biodiversity can be broken down into three levels: intraspecific (the genetic diversity of a cultivated species, for example), interspecific (the number of species present in a given environment) and environmental or ecosystem diversity.

Bioeconomy According to the Organisation for Economic Co-operation and Development (OECD), the bioeconomy refers to the set of economic activities relating to the invention, development, production and use of biological products and processes. A "circular" bioeconomy means it does not require—or requires very little—input from non-renewable sources. Because the concept is new, its definition is constantly evolving.

M. Guillou, G. Matheron, *The World's Challenge*,
DOI 10.1007/978-94-017-8569-3, © Éditions Quæ, 2014

Bioenergy The various ways in which biomass is used to produce energy. Types of bioenergy include biodiesel and bioethanol (used in road and air transport) and solid or liquid biofuels and biogas (used to produce heat and electricity).

Biofuel Biomass-derived products that can be substituted for fossil fuels. First-generation biofuels use plant storage organs, either directly, as with oleaginous plant oils, or after fermentation, such as cereal starches after hydrolysis and cane or beet sugars. Second-generation biofuels are derived from biomass waste and residues or whole plants.

Biomass Organic matter from plants or animals.

Bioproducts Biomass-derived products that include biomaterials (e.g. biodegradable polymers, fibres and composite materials, paper and cardboard), biochemicals (e.g. surfactants, solvents, lubricants and cosmetics) and stabilised living agents (starters).

Biosphere From an ecological standpoint, the Earth is made up of the hydrosphere (oceans and other aquatic environments), the lithosphere (the uppermost layer of Earth) and the atmosphere (the gaseous environment surrounding the Earth). The biosphere is the thin, superficial layer of our planet where life can exist, consisting of the hydrosphere, the lithosphere and the lowest part of the atmosphere.

Biotechnology Biotechnology is the application of science and technology to living organisms to create biological tools and processes. Each sector comprises various disciplines, methods, know-how, tools and processes. The various biotechnologies can be viewed as a spectrum. White biotechnology is applied to industrial uses (excluding medical applications); green refers to the processing of plants and therefore to agriculture; red applies to medical uses; blue refers to marine applications and yellow to environmental protection. More specifically, white biotechnologies mobilise biological processes for the manufacture of chemical substances using the breakdown and transformation of living or non-living matter by enzymes or bacteria. They are used in conjunction with classic chemical processes to meet both economic and environmental objectives.

BMI Body mass index. Calculation of a person's weight, expressed in kilograms, divided by the square of his or her height, expressed in metres. A normal BMI falls between 18.5 and 24.9; a BMI below 18.5 indicates a person is underweight while a BMI above 30 indicates a person is overweight.

Cellulose Main constituent of plant cell walls. It is the most abundant organic component on Earth, making up 50 % of all plant biomass.

Club of Rome Think tank that brings together scientists, economists and officials from national and international institutions, as well as industrial operators from 53 countries to deal with complex issues that all countries, whether developing or industrialised, must face. It takes its name from its first meeting, held in Rome in 1968. Its reputation as a forerunner is due to the fact that the notions of sustainable development and carbon footprints were first introduced here. It earned worldwide recognition after publishing its first report in 1972, *The Limits to Growth*, also known as the Meadows Report.

Collective scientific expert report. It aims to assess the state of scientific knowledge on a given issue. INRA has coordinated collective expertise reports since 2002.

Contingent valuation Method that aims to assess, via direct surveys, how much people would be willing to pay for a good or service with a specific characteristic (taste, environmental) compared to a good or service without that characteristic. The sum consumers would pay is their willingness-to-pay (WTP).

DDT Abbreviation for dichlorodiphenyltrichloroethane, the first "modern" insecticide developed at the beginning of the Second World War. It was successfully used against mosquitoes carrying malaria and typhus, other disease-carrying insects and as an agricultural pesticide. It has been banned in numerous countries since the 1970s.

Decarbonisation Term referring to the replacement of fossil carbon by renewable biomass carbon.

Decision support tool Methods, algorithms and software that enable the user to choose the best solution from various possible solutions. Decision support tools provide relevant answers to diverse problems by offering various possible choices, such as the implantation of industrial sites, a strategy for cleaning up a lake, investment portfolio creation, etc. They can also help with carrying out a diagnosis and facilitating strategic or operational decision-making in an imprecise or uncertain environment.

Degenerative disease Non-communicable disease often linked to poor lifestyle habits (poor diet, lack of exercise, alcohol or medication abuse). Examples include type 2 diabetes, heart diseases, and certain cancers and joint diseases.

Developed (industrialised)/emerging market/developing/transition countries No single definition exists. Generally speaking, an emerging market country is one where per capita income is below those of developed countries (OECD members), but where the growth rate exceeds the world average. Brazil, Russia, India and China are the main emerging market countries (BRICs). The term "developing" applies to those countries which are not developed. The World Bank has a more narrow definition, applying the term to countries with a low or middle per capita gross domestic product (GDP). Also included are five economies, which, despite having higher incomes, are classed as developing due to their economic structures or national political positioning. Of developing countries, those which are the least developed are those with low per capita income, lagging in human development and economically weak. Lastly, transition countries are those nations which are emerging from a centrally-controlled economy and moving towards a market economy.

Dopamine (circuits) Systems involved in brain and nervous system functions, which need dopamine, a neurotransmitter, to work.

Doubly green revolution Essentially, this term refers to the need to pursue increasing agricultural production and crop yields while respecting the environment and natural resources. More specifically, it refers to productive and environment-friendly agricultural practices and systems that make best use of the natural functions of cultivated ecosystems.

Dry matter Residue obtained when the water from a product is removed.

Ecological intensification In simple terms, ecological intensification is used to describe agricultural practices and systems that best use the services provided by nature for their production in order to save fossil fuels, reduce the use of mineral fertilisers and pesticides, and minimise environmental impacts.

Ecosystem In ecology, ecosystems are units or communities consisting of living organisms and their environment, called a biotope, which includes biological, geological, edaphic, hydrological and climatic factors. The elements within an ecosystem exchange energy and matter, allowing life to develop.

Ecosystem services There are four major divisions of ecosystem services: (1) supporting services that ensure the basic ecosystem functions, such as the carbon cycle; (2) provisioning services (food, water, etc.); (3) regulating services, from regulation of flood risks by plants to plant pollination; and (4) cultural services, including the heritage value of a landscape or protected species.

Ecotone Ecological transition area between two different ecosystems, such as forest and savanna or alluvial plains and high plains.

Environmental services A sub-division of ecosystem services relating to the environment, including water purification by ecosystems, biodiversity protection or carbon storage in the soil.

Enzyme Protein produced by a living organism which increases the rate—up to millions of times faster—of biochemical reactions inside or outside cells without altering the established balance. Enzymes are only needed in low quantities and remain intact at the end of the reaction; they are biocatalysts.

Eukaryotes Single or multi-celled organisms characterised by a nucleus and the presence of mitochondria within their cells. Eukaryotes are common to four kingdoms: protists (single-celled plant and animal organisms), fungi, plants and animals.

FAO United Nations Food and Agricultural Organization (191 Member Nations, two associate members and one member organisation, the European Union). Its objective is to ensure that all people have regular access to sufficient food of good quality so that they can lead healthy and active lives via agricultural production and agri-food product consumption.

Flavour The taste and smell of food perceived by the consumer.

Food calorie 1 food calorie is equal to 1,000 thermal calories (4,200 J).

Food safety risks There are generally two types of food safety risks: health risks associated with consuming foods that have been contaminated in some way which pose a danger to the health of the consumer (infection, poisoning, etc.) and nutritional risk due to malnutrition.

Food security The World Food Summit of 1996 defined food security as existing "when all people at all times have access to sufficient, safe, nutritious food to maintain a healthy and active life". Commonly, the concept of food security is defined as including both physical and economic access to food that meets people's dietary needs and corresponds to their food preferences.

FRB Foundation for Research on Biodiversity. Originally an initiative led by the French ministers for Research and Ecology, the FRB was created in France in 2008 by eight public research institutions and organisations. The FRB's objective is to promote research activities on biodiversity in collaboration with major players in the industry. It encourages innovation, develops and supports projects, disseminates knowledge and builds expertise.

Fresh dairy products Refrigerated dairy products which must be consumed within a few days and which must be stored at cold temperatures (between 0 and 6 °C).

Functional foods According to the International Life Sciences Institute (ILSI), functional foods are those that have beneficial effects on body functions that go beyond adequate nutritional effects, contributing to an improved state of health and well-being and/or reducing the risk of disease.

Functional trait (or characteristic) Morphological, physiological or even behavioural quality of a living organism that plays a vital role in its growth, functioning and reproduction.

G8 Group of eight countries: Canada, France, Germany, Italy, Japan, Russia, the United Kingdom and the United States. The heads of state or government of each of these countries and the presidents of the European Commission and European Council meet yearly at the G8 summit.

G20 Group of 19 countries: Argentina, Australia, Brazil, Canada, China, France, Germany, India, Indonesia, Italy, Japan, Mexico, Saudi Arabia, South Africa, South Korea, Turkey, the United Kingdom and the United States, plus the European Union. The group, created in 1999, brings together heads of state and government, ministers and central bank leaders. Its objective is to promote concerted decision making and international cooperation within a framework of expanded dialogue which takes into consideration the growing economic weight of emerging market countries.

Genome Genetic matter of an individual or species, carried mainly—but not exclusively—on the chromosomes in cell nuclei. Genomes contain all coding DNA sequences (i.e., those that code for proteins) and non-coding sequences (those that do not code for proteins).

Genomics The field of genomics can be divided into two branches: the study of genome structure (study of DNA sequencing) and the study of gene expression (building of protein molecules).

Genotype Genotype can refer either to an individual's collection of genes (genetic code) or to the two alleles inherited for a particular gene. In a strict sense, genotype designates the code of a restricted number of genes. For example, AB genotype for the coding gene for the ABO blood group, located on human chromosome 9 and having 3 possible alleles, A, B and O.

Glucose One of the main simple sugars, which is directly assimilated by the body.

Glycemic index Ranking of different foods containing carbohydrates according to the degree and speed at which blood glucose level rises after their absorption (glycemia). The glycemic index ranks carbohydrates on a scale from 0 to 100, with 100 being pure glucose.

Green revolution Term referring to an increase in agricultural yields from 1960 to 1990, made possible by improvements in crop varieties, use of irrigation, mineral fertilisers and pesticides, coupled with investments in infrastructure (storage, transport, processing) and in research and development. Drawbacks often include increased stress on the environment and natural resources.

Greenhouse gases (GHGs) Used to describe gases which absorb infrared rays reflected from the Earth's surface and which contribute to the greenhouse effect. This is similar to what happens inside a greenhouse, where infrared rays cannot escape, energy is retained and the temperature increases. The main GHGs are water vapour (H_2O), carbon dioxide (CO_2), methane (CH_4), nitrous oxide (N_2O), and ozone (O_3). This is a perfectly natural phenomenon that is essential for life on Earth. However, problems arise when the concentration of GHGs in the atmosphere increases rapidly due to human activity.

Halo effect The halo effect is a cognitive bias by which a decision or behaviour in a given situation is based on a false perception of available information. When applied to agri-food products, it implies that the name of the food product, its brand or an unproven claim about the product have greater impact on expectations in terms of nutritional quality and taste than the nutritional description of the contents.

Hay Method of conserving fodder in dry form, obtained by drying the grass completely in the sun (natural drying) or finishing the process artificially with warm air (barn drying).

Hedonic pricing The hedonic pricing method is used to estimate economic values for a good or service based on its various characteristics.

Hemicellulose A component of plant cell walls.

Hexose, pentose Two simple sugars that cannot be hydrolysed (i.e. cannot be broken down in water), which form crystals and have six and five carbon atoms, respectively.

Lignin One of the main components of wood, making it rigid, water resistant and slow to decompose. Lignin is the second-most renewable biopolymer on Earth, after cellulose.

Low water level Period of the year when the flow of water reaches its lowest point.

Macronutrients, micronutrients Macronutrients include carbohydrates, fats and proteins. Micronutrients are organic substances present in food which the body requires in small amounts, such as trace elements and vitamins.

Malnutrition Poor diet due to an excess of calories (overnourishment) or insufficient calories (undernourishment) or other calorie imbalance.

Malthus (Thomas Robert, 1766–1834) British reverend and economist, famous for his pessimistic views of demographic growth and agricultural productivity. The adjective "Malthusian" is derived from his name, as well as a doctrine referred to as "Malthusianism". Both refer to the notion that land is a rare commodity and that the planet cannot be fed without some form of population control.

Meadows Report Report commissioned in 1970 and published in 1972 by the Club of Rome, entitled "The Limits to Growth". For the first time, the benefits of growth were questioned by researchers at the Massachusetts Institute of Technology (MIT) in light of a forecasted energy shortage and the negative impact of industrial progress on the environment.

Millennium Ecosystem Assessment (MA) Ecosystem assessment programme launched in 2001 by the United Nations.

Molecular marker Identifiable genetic variation at a given location on the genome. Analysing these variations (genotyping) falls within the field of molecular biology. The uniform presence of markers in large numbers on the genome can be used to predict the presence of certain alleles of genes located nearby and associated with parameters of interest in animal and plant breeding.

Mycotoxins Secondary toxic metabolites produced by certain moulds or fungi on plants in the field or during storage. More than 300 mycotoxins have been identified, some thirty of which are seriously toxic to humans or animals.

Northern countries/Southern countries Generally speaking, developed countries are also referred to as "Northern countries", while developing and less developed countries are referred to as "Southern countries".

Nutrition education According to the French Institute for Nutrition (IFN), nutrition education is an educational process of teaching nutritional knowledge that aims to make changes in food behaviour adapted to an individual's specific needs.

Organic matter Organic matter differs from mineral matter by its ability to decompose. It can come from plants or animals and is an important component of soil.

Percolation In the context of this book, the downward movement of water through the soil under the effect of gravity.

Perishable foodstuff All food products have a limited lifetime, at the end of which they may no longer be consumed. Perishable foodstuffs refer to those which may only be kept for short periods of times, such as root and other vegetables, tubers, fruits, meats and fish, as opposed to those which may be stored for a long period of time, e.g. cereals.

Pesticide Pesticides encompass three product families that aim to protect crops from weeds (herbicides), microscopic fungi (fungicides), and insects (insecticides). This definition makes no distinction between pesticides and plant protection products. EU regulations allow for a wider meaning, defining pesticides as chemical substances that are used to repel, kill or regulate the growth of undesirable organisms, plants or animals that damage food products, crops, wood and wood products, in addition to food for animals.

Phenotype Set of characteristics observed in a living organism. A phenotype results from the interaction of its genotype with the environment. The term is often restricted to refer to a single characteristic.

Photosynthesis Process by which plants and certain bacteria, drawing their energy from the sun, use carbon dioxide, water and mineral salts to synthesise organic matter.

Post-harvest losses Losses occurring during transport, drying, storage, packaging or getting products to market. They may be mechanical (impacts, damage) or biological (mould, insects, predators).

Processing With regards to crops, livestock or fishing, processing refers to the various stages of change that agricultural commodities undergo. Processing activities are divided into two levels. First processing generally deals with products that have a low added value (e.g., flour) while products that undergo a second processing have higher added value (e.g. bread, ready-to-eat meals, etc.).

Productivity Generally speaking, in agriculture, productivity measures the efficiency of the transformation process of one or more factors of production (land, water, labour, capital, fertiliser, etc.) into a result. When used in reference to one factor, the term "partial productivity" is used, while the term "total or overall productivity" is used to refer to the whole set of factors involved. The partial productivity is the ratio of production to the surface area, i.e. yield.

Public goods Public goods—and by extension, services—are those with non-rival consumption (the use of that good is not limited to one single entity), and having non-exclusive benefits (the good may be freely accessed). There are global public goods and services, for which the usefulness of production by local or national players extends beyond national borders (e.g., climate regulation), and local public goods, the benefits of which remain confined to a more restricted area (e.g., bee pollination). As opposed to private goods and services, which are rival and exclusive, there is often an insufficient supply of public goods and services due to the nature of the single market. The public sector must then intervene to restore optimum levels.

REACH (regulation) REACH is the European Community regulation dealing with the Registration, Evaluation, Authorisation and Restriction of Chemical substances. The law, which entered into force on 1 June 2007, improves EU regulation of chemical substances and aims to improve human health and environmental protection while maintaining competitiveness and enhance innovation in the European chemical industry.

Resilience Generally speaking, the ability of a system to return to normal after a disturbance or stress.

Revealed preference Method of measuring consumer preferences by observing their purchasing behaviour, rather than questioning them beforehand by suggesting various alternative goods.

Saccharification Biochemical process of breaking a complex carbohydrate such as cellulose or starch into simple sugars like fructose and glucose.

Sensory education According to the French Institute for Nutrition (IFN), sensory education is an educational process meant to widen sensory perception of food. Sensory appreciation helps individuals develop a deeper appreciation for a variety of food.

Silage Method of conserving fodder via lactic fermentation. By way of analogy, the process is similar to making sauerkraut. The term silage is also used for the product obtained: grass silage, maize silage, etc.

Soil fertility The capacity of a soil to produce a crop. Soil fertility is inextricably linked to climatic conditions, which affect production potential. It is also influenced by previous crops grown and methods of production used: crop species and sequences, tillage, fertilisers, etc.

Spirulina Edible filamentous cyanobacteria which are grown industrially due to their high nutritional value. Spirulina is commonly known as “blue-green algae”.

Trans fatty acids (trans fat) Natural or artificial hydrogenated unsaturated fatty acids. Total daily calorie intake of over 2 % of artificial trans fats has been linked to higher risks of heart disease because it increases levels of “bad” (LDL) cholesterol in the blood.

Washington Consensus The Washington Consensus was named after an article by American economist John Williamson. In 1989, Williamson outlined ten policy recommendations that developing countries should adopt to reduce debt levels. They mainly emphasised controlling public spending and deregulating the economy. For many years, loans granted by international institutions headquartered in Washington, i.e. the World Bank and the International Monetary Fund (IMF), were based on the adoption of these policies.

Weed Undesirable herbaceous or woody plant.

World Food Programme (WFP) The largest humanitarian agency fighting hunger in the world, entirely financed by voluntary contributions. The WFP, an integral part of the United Nations organisational structure, fights hunger throughout the world by intervening in emergency situations (food crises and post-crisis support) and in certain chronic food shortage situations. It also provides aid to help countries in need reduce hunger by purchasing food, handling resource transfers, etc.

The manufacturer's authorised representative in the EU is Springer Nature Customer Service Centre GmbH, Europaplatz 3, 69115 Heidelberg, Germany. If you have any concerns regarding our products, please contact ProductSafety@springernature.com

Printed and bound by CPI Group (UK) Ltd, Croydon, CR0 4YY
30/09/2026
02225445-0002